Energy, Climate and the Environment

Series editor
David Elliott
The Open University
Milton Keynes, UK

Aim of the Series

The aim of this series is to provide texts which lay out the technical, environmental and political issues relating to proposed policies for responding to climate change. The focus is not primarily on the science of climate change, or on the technological detail, although there will be accounts of this, to aid assessment of the viability of various options. However, the main focus is the policy conflicts over which strategy to pursue. The series adopts a critical approach and attempts to identify flaws in emerging policies, propositions and assertions. In particular, it seeks to illuminate counter-intuitive assessments, conclusions and new perspectives. The intention is not simply to map the debates, but to explore their structure, their underlying assumptions and their limitations. The books in this series are incisive and authoritative sources of critical analysis and commentary, clearly indicating the divergent views that have emerged whilst also identifying the shortcomings of such views. The series does not simply provide an overview, but also offers policy prescriptions.

More information about this series at
http://www.springer.com/series/14966

Geoffrey Wood · Keith Baker
Editors

A Critical Review of Scottish Renewable and Low Carbon Energy Policy

palgrave
macmillan

Editors
Geoffrey Wood
School of Law
University of Stirling
Stirling, UK

Keith Baker
School of Engineering and the Built Environment
Glasgow Caledonian University
Glasgow, UK

Energy, Climate and the Environment
ISBN 978-3-319-56897-3 ISBN 978-3-319-56898-0 (eBook)
DOI 10.1007/978-3-319-56898-0

Library of Congress Control Number: 2017940366

Printed on acid-free paper

This Palgrave Macmillan imprint is published by Springer Nature
The registered company is Springer International Publishing AG
The registered company address is: Gewerbestrasse 11, 6330 Cham, Switzerland

To Mark Rylatt, in lieu of beer, and to Yana, Katya and Ira, my motivation.

Series Editor Preface

Concerns about the potential environmental, social and economic impacts of climate change have led to a major international debate over what could and should be done to reduce the emissions of greenhouse gases. There is still a scientific debate over the likely *scale* of the severity of climate change and the complex interactions between human activities and climate systems, but global average temperatures have risen, and the cause is almost certainly the observed build-up of atmospheric greenhouse gases.

Whatever we now do, there will have to be a lot of social and economic adaptation to climate change—preparing for increased flooding and other climate related problems. However, the more fundamental response is to try to reduce or avoid the human activities that are causing climate change. That means, primarily, trying to reduce or eliminate the emission of greenhouse gases from the combustion of fossil fuels. Given that around 80% of the energy used in the world at present comes from these sources, this will be a major technological, economic and political undertaking. It will involve reducing demand for energy (via lifestyle choice changes and policies enabling such choices to be made), producing and using whatever energy we still need more efficiently (getting more from less), and supplying the reduced amount of energy from non-fossil sources (basically switching over to renewables and/or nuclear power).

Each of these options opens up a range of social, economic and environmental issues. Industrial society and modern consumer cultures have been based on the ever-expanding use of fossil fuels, so the changes required will inevitably be challenging. Perhaps equally inevitable are disagreements and conflicts over the merits and demerits of the various options and in relation to strategies and policies for pursuing them. These conflicts and associated debates sometimes concern technical issues, but there are usually also underlying political and ideological commitments and agendas which shape, or at least colour, the ostensibly technical debates. In particular, at times, technical assertions can be used to buttress specific policy frameworks in ways which subsequently prove to be flawed.

The aim of this series is to provide texts which lay out the technical, environmental and political issues relating to the various proposed policies for responding to climate change. The focus is not primarily on the science of climate change, or on the technological detail, although there will be accounts of the state of the art, to aid assessment of the viability of the various options. However, the main focus is the policy conflicts over which strategy to pursue. The series adopts a critical approach and attempts to identify flaws in emerging policies, propositions and assertions. In particular, it seeks to illuminate counter-intuitive assessments, conclusions and new perspectives.

The present text is no exception in exploring the ambitious renewable energy programme underway in Scotland. Scotland is aiming to expand the output of renewables, so that they generate the annual equivalent of all its electricity consumption by 2020. At the time of writing, it has reached over 60%, well ahead of most other countries in the world, apart from those with large existing hydro capacities.

Scotland remains part of the UK, at least for the present, but it has a devolved government, led by the Scottish National Party (SNP), and its polices on energy are clearly different from those of the Westminster Government, including its opposition to new nuclear. The UK's vote in 2016 in favour of leaving the EU may lead Scotland to seek another referendum on independence from the UK, since the EU referendum showed a significant majority of Scots wanted to stay in the EU. In which case, its energy policy could diverge even more. That is speculative, but what is no longer speculative is Scotland's ability to install and operate increasing amounts of renewable capacity.

That is not to say there are no critics of Scotland's renewable energy programme; some depict it as foolish or at least of limited value and high cost. Some of the criticisms are simply due to disbelief that renewables such as wind energy (now the dominant renewable in Scotland) can work effectively on a very large scale, without massive backup. Certainly, balancing issues are coming to the fore. Some critics also resent the SNPs opposition to nuclear power, which they see as a vital component of a balanced system. The chapter on nuclear in this book reflects that view and suggests a rethink may be in order, or at least full consideration of what the phase-out of the two Scottish nuclear plants would imply. Much of the rest of the book, in effect, offers some ideas for new areas of development, in addition to wind power (which seems likely to remain the main option), with chapters on marine energy (wave and tidal), community energy projects and energy efficiency, including heating issues, a key area for the future, so far poorly addressed in Scotland, as in the UK.

Clearly, there are many options and some urgent policy and development issues to be faced, and this book offers a guide to how a devolved, and possibly independent, Scottish Government could address them. Not all of the issues are addressed fully in this book. Although it sets the wider scene, it focuses on non-fossil energy options: renewables, nuclear power and energy efficiency. So it does not cover fossil fuel issues in any detail, apart from CCS, and only delves briefly into transport issues, focusing on user behaviour rather than technology. Nevertheless, it still provides a timely and critical account of the potential and likely problems of what many see as a brave attempt to accelerate renewables, so that they can meet most energy needs, while also allowing for continued export of electricity.

Milton Keynes, UK David Elliott

Acknowledgements

The editors would like to thank all those who contributed chapters to this book and all those who gave their support in other ways, including: Prof. Dave Elliot for his unwavering support throughout and Chloe Fitzsimmons and Rachael Ballard at Palgrave Macmillan; Prof. Rohinton Emmanuel, Dr. Mark Phillipson and Dr. Craig Thomson, Glasgow Caledonian University; Ron Mould, Renfrewshire Council; Scott Restrick, Energy Action Scotland; Prof. Sue Roaf, Heriot-Watt University; Prof. Peter Cameron, Stephen Dow and Prof. Colin Reid, Dundee University. Thanks also to Sarah Garden who kindly provided the indexing and to Academic Proofreading and Editing Services (https://academicproofreadingandeditingservices.wordpress.com/2015/07/21/38/) for the final proofreading and editing of the manuscript.

Contents

Part II The Challenges Ahead

Editors and Contributors

About the Editors

Dr. Geoffrey Wood is Teaching Fellow in International Energy Law and Policy at Stirling University. An energy and environmental law, policy and regulatory specialist, his work focuses on the development of legal and governance frameworks for sustainable energy transitions to optimise policy delivery in terms of environmental, social, economic, political and technological outcomes. He has published over 30 articles, books, chapters and other publications. Currently also working as a Researcher and Contributing Lecturer at Dundee University, Geoffrey has previously held research, lecturing and consultancy positions for various organisations, including the Scottish Government and the Offshore Renewables Institute.

Dr. Keith Baker is a Researcher at the School of Engineering and the Built Environment, Glasgow Caledonian University and a co-founder of the Initiative for Carbon Accounting (ICARB). He specialises in energy in the built environment, with a particular focus on fuel poverty, energy efficiency, and sustainable building materials.

Contributors

Keith Baker School of Engineering and the Built Environment, Glasgow Caledonian University, Glasgow, UK

Anna L. Berka Department of Forest Sciences, University of Helsinki, Helsinki, Finland

Jelte Harnmeijer University of Edinburgh, Edinburgh, Scotland, UK; James Hutton Institute, Dundee, Scotland, UK

Raphael J. Heffron Queen Mary University of London, London, UK

William J. Nuttall The Open University, Milton Keynes, UK

Bill Slee James Hutton Institute, Dundee, Scotland, UK

Alan Taylor University of Dundee, Dundee, UK

David Toke Department of Politics and International Relations, University of Aberdeen, Aberdeen, UK

Geoffrey Wood School of Law, University of Stirling, Stirling, UK

Abbreviations

AGR	Advanced Gas-Cooled Reactor
ASHP	Air Source Heat Pumps
BREDEM	Building Research Establishment's Domestic Energy Model
CARES	Community and Renewable Energy
CCL	Climate Change Levy
CCS	Carbon Capture and Storage
CE	Crown Estate
CERT	Carbon Emissions Reduction Target
CES	Community Energy Scotland
CESP	Community Energy Saving Programme
CfD FiT	Contract for Difference Feed-in Tariff
CHP	Combined Heat and Power
CO_2	Carbon Dioxide
CoE	Cost of Energy
DCC	Data Communications Company
DECC	Department for Energy and Climate Change (closed as of 2016 with the energy component merged into the new Department for Business, Energy and Industrial Strategy)
DEMScot2	Domestic Energy Model for Scotland
DETI	Department of Enterprise, Trade and Investment (Northern Ireland)
DNV	Det Norske Veritas
EAP	Energy Assistance Package
ECI	Energy Consumption Indicators
ECO	Energy Company Obligation
EDF	Electricitié de France

EES	Energy Efficiency Commitment
EIA	Environmental Impact Assessment
EIS	Enterprise Investment Scheme
EMEC	European Marine Energy Centre
EMR	Electricity Market Reform
ENSG	Electricity Networks Strategy Group
ENTSO-E	European Network Transmission System Operator—Electricity
EPSRC	Engineering and Physical Sciences Research Council
ERDF	European Regional Development Fund
EST	Energy Saving Trust
ETI	Energy Technologies Institute
ETP	Energy Technology Partnership
EU	European Union
FiT	Feed-in Tariff
GB	Great Britain
GHG	Greenhouse Gas
GSHP	Ground Source Heat Pumps
HICEC	Highlands and Islands Community Energy Company
HIE	Highlands and Islands Enterprise
HIS	Home Insulation Scheme
IEA	International Energy Agency
IRENA	International Renewable Energy Association
kV	Kilovolt
LCF	Levy Control Framework
LCoE	Levelised Cost of Energy
LEED	Leadership in Energy and Environmental Design
LRRG	Land Reform Review Group
MEA	Marine Energy Accelerator
MEAD	Marine Energy Array Demonstrator
MEG	Marine Energy Group
MESAT	Marine Energy Supporting Array Technologies
MFA	Marine Farm Accelerator
MRCF	Marine Renewables Commercialisation Fund
MRPF	Marine Renewables Proving Fund
NERC	Natural Environment Research Council
NFFO	Non-Fossil Fuel Obligation
NGO	Non-Governmental Organisation
NIE	Northern Ireland Executive
NIRO	Northern Ireland Renewables Obligation
NPF	National Planning Framework (Scotland)
O&G	Oil and Gas
OECD	Organisation for Economic Cooperation and Development

OES	Ocean Energy System
OFGEM	Office for Gas and Electricity Markets
ORE	Offshore Renewable Energy Catapult
OWCRTF	Offshore Wind Cost Reduction Task Force
PCA	Personal Carbon Allowances
PMSS	Project Management Support Services Limited
PPA	Power Purchase Agreement
PV	Photovoltaic (also called Solar Photovoltaic)
R&D	Research and Demonstration
RDF	Rural Development Fund (Scotland)
rdSAP	Reduced Data Standard Assessment Procedure
RECAI	Renewable Energy Country Attractiveness Index
RECO	Residential Energy Conservation Ordinance
REIF	Renewable Energy Investment Fund
RES-E	Electricity Generated from Renewable Energy Sources
RET	Renewable Electricity Technology
RHI	Renewable Heat Incentive
RO	Renewables Obligation
ROC	Renewables Obligation Certificate
ROS	Renewables Obligation Scotland
rUK	Rest of the UK (In the event Scotland becomes independent from the UK)
SAP	Standard Assessment Procedure
SBEM	Simplified Building Energy Model
SEEP	Scotland's Energy Efficiency Programme
SEIS	Seed Enterprise Investment Scheme
SEP	Smart Energy Profile
SHCS	Scottish House Conditions Survey
SMRU	Scottish Marine Research Unit
SNH	Scottish Natural Heritage
SNP	Scottish National Party
SO	System Operator
SPS	Strategy and Policy Statement
SROC	Scottish Renewable Obligation Certificate
TRL	Technology Readiness Level
TSB	Technology Strategy Board
UK	United Kingdom
UN	United Nations
US	United States (or United States of America)
WATERS	Wave and Tidal Energy: Research, Development and Demonstration Support
WEC	Wave Energy Converter
WHO	World Health Organisation

Note on Units

Power units	The power using or generating capacity of devices is measured in watts, or more usually kilowatts (kW) (1 kW = 1,000 W). Larger units are megawatts (MW) (1,000 kW), gigawatts (GW) (1,000 MW) and terawatts (TW) (1,000 GW).
Energy units	The kilowatt-hour (kWh) is the standard unit by which electricity is sold—1 kWh is the energy produced/consumed when a 1 kW rated generator/energy-consuming device runs for 1 h. A megawatt-hour (MWh) is 1,000 kWh. Similarly, 1,000 MWh = 1 GWh and so on.

List of Figures

List of Tables

PART I

Opportunities and Limitations

CHAPTER 1

Introduction: Aye. Naw. Mibbe.

Geoffrey Wood and Keith Baker

1.1 Introduction

Aye. Naw. Mibbe.[1] Three words that perfectly capture the astonishing and unparalleled series of recent events in Scottish politics held in a Death Star-like grip by the issues of independence, devolution, governance and the right for Scotland to gain increasing control over its own affairs. *Will we? Should we? Could we?* Vote for independence? Further devolution? Retain the existing settlement? Not as simple as they first seemed, these questions opened up the very nature of the existing and future relationship between Scotland and the United Kingdom (UK). In short, these interdependent, complex issues can be termed the 'independence debate', and it is one that shows no signs of resolution or fading away.

Indeed, it has become one of the defining points of the new millennium in Scotland and the wider UK, from the resurgence of the

G. Wood (✉)
School of Law, University of Stirling, Stirling, UK
e-mail: geoffrey.wood@stir.ac.uk

K. Baker
School of Engineering and the Built Environment,
Glasgow Caledonian University, Glasgow, UK
e-mail: keith.baker@gcu.ac.uk

G. Wood and K. Baker (eds.), *A Critical Review of Scottish Renewable and Low Carbon Energy Policy*, Energy, Climate and the Environment,
DOI 10.1007/978-3-319-56898-0_1

independence debate with the Scottish National Party (SNP) winning power in 2007 and every election north of the border since, to the 2014 independence referendum and the prematurely proclaimed demise of the idea, to an unexpected and painfully unprepared Brexit, and now on towards a second referendum and calls for a fundamentally altered UK, particularly in terms of how it is perceived and operates on the global stage.

This book, with its unique focus on Scottish renewable and low-carbon energy policy and practice that critically reviews the opportunities and challenges going forward, both real and plausible, is a contribution to this ongoing debate. It can be argued that the contemporary origins of the independence debate appeared with the discovery of commercially viable hydrocarbon reserves in the Scottish North Sea and the emergence of calls for independence culminating in the failed 1978 devolution vote. It is therefore right that renewable and low-carbon energy should play centre stage in the debate being carried out now, in a world facing the threat of climate change, energy security concerns and the potential economic and political gains from developing domestic and export capabilities and resultant job growth. The world is moving towards a sustainable low-carbon transition, with many arguing that renewable energy has already reached a tipping point[2] as sustainable technologies become economically viable, investment levels soar year-on-year and novel technologies become increasingly embedded within the political, industry and public consciousness and as what appeared only recently to be virtually impossible becomes more and more a reality.

For now, the people of Scotland have decided to stay as part of the UK, although by a slimmer majority than many on both sides of the debate thought possible. Despite the UK overall voting to leave the European Union (EU), Scotland (along with Northern Ireland and Gibraltar) also voted overwhelmingly to remain within the EU in contrast to England and Wales. Once again, the independence debate has arisen with a vengeance on the back of the EU referendum and the attitude of a UK government favouring a hard Brexit. And once again, further devolution or independence is firmly back on the political table. Although future developments are invariably difficult to determine in advance, it may seem strange to publish a book on Scottish energy policy before knowing the outcome of the ongoing debate; however, its timing is deliberate for a number of reasons. Let us explain why.

First of all, it allows us to present opinion that is agnostic of the results whilst being cognisant of the evidence of how things are likely to change, or not, under either outcome. Secondly, either outcome will still lead to at least several years of negotiations over policies that will affect the populations of both sides of the border, with energy policy being a key and potentially highly polarising example. So, it would be unwise to assume Scotland's energy future will be dramatically different. But finally, and most importantly, and as discussed in the chapters to follow, the seeds of that energy future were sown long before the SNP took the power needed to enact its mandate of putting the question to the people, and the implications of long-term energy policies set years before the debate will be felt for many years to come and, to a greater or lesser extent, regardless of the outcome.

Ask any Scot what they think the country's biggest energy issues are and their likely answer will include the oil industry, wind farms or nuclear power. The majority of Scots oppose nuclear power, like wind farms, and would like a bigger share of the income from the oil industry, and so it came as no surprise when these became big political footballs in the lead up to the referendum. However, the outcome of the referendum will have little bearing on these issues—the development of any new wind farms or nuclear plants is already largely in the hands of the Scottish Government and in the long term the country will still need to wean itself off oil—so any future government would be unlikely to change direction on them without a significant shift in public opinion. As such they have served as convenient distractions from other more contentious games being played out in the run-up to the referendum, especially where their devolutionary status means they can be held hostages to fortune, and neither side can claim to be innocent of this.

For this reason, the emphasis of this book is very much on the *critical*, and so we have brought together contributions from experts from across the field of energy policy research and encouraged them to pull no punches in their analyses. Supporters of either side will find little solace here. Whilst the Scottish Government has racked up numerous achievements in decarbonising the energy sector, it has shied away from other opportunities to tackle both supply and demand, either directly or through influencing Westminster. And whilst there is little doubt that it has pursued a far more progressive strategy than that inflicted on England and Wales, this gulf between Scotland and the rest of the UK is significantly widened by the scale of Westminster's failures.

Our planet does not much care for the outcome of a decision that will affect the nature of an arbitrary line drawn across an island, and climate change will not somehow magically pause whilst five and a half million people wait the outcome of the next round of the debate. However, if the political will that has been brought to bear on determining the future of those people could be harnessed for energy and climate change policy then both sides would have an awful lot more to shout about to the other seven billion.

Scotland can lay claim to some of the greatest renewable energy resources in Europe, and for a country of its size its potential is enviable on a global level. Yet it also suffers from the temptations of accessible oil, gas and coal—three profitable industries that will need to be largely or completely eradicated in the cause of mitigating climate change—and then there is the problem of managing demand, and ensuring it can be met without the need for imports of non-renewable energy. A successful transition to one hundred per cent renewable energy would be the key practical outcome of the targets put to paper in the Climate Change (Scotland) Act 2009, however it remains to be seen if the policies being put in place to implement it will be sufficient to turn those world-leading aspirations into world-leading achievements, and it should come as no surprise that many of the analyses offered in the chapters to follow cast doubt on whether those targets will be met.

As editors, we are indebted to the time and effort put in by all the contributors to this book, and we hope that readers will find it an informative and challenging journey through Scottish energy policy.

1.2 Outline of the Book

The rest of the book sets out these issues in more detail. Part One sets out the current context of renewable and low-carbon energy policy and practice in Scotland in terms of the opportunities and limitations going forward. Chapter 2 examines the development of Scottish renewable electricity policy under devolution from 1997 prior to the independence referendum and the Smith Commission. It specifically focuses on the distribution of powers and the divergence in policy and practice between the UK and Scotland, and the implications for large-scale renewable technology deployment which contributes most to meeting targets. This is relevant given the energy sector is once again entering a new phase of radical reform via the electricity market reform process. Two key points are made here. Devolution has resulted in significant benefits for the Scottish Government in gaining legislative competence and the legitimised capacity to influence UK policy. However, it has not resulted in a clear demarcation of powers between Westminster and Holyrood, leading to largely individualistic, piecemeal and arbitrary arrangements in terms of what is reserved and devolved. This reopens the debate over whether a comprehensive and cohesive set of devolved powers over renewables would be advantageous not just to Scotland but the UK overall.

This is followed by an examination of the hopes and challenges for community-scale renewables in Scotland in Chap. 3. The development of community renewables in Scotland is interwoven with a range of post-devolution Scottish policies relating to community ownership of natural resources and community empowerment, and facilitated and hindered to varying degrees by a dynamic and uncertain energy policy landscape at the UK and EU levels. Whilst community energy production represents only about 4 per cent of Scottish onshore renewable generation, it can provide highly important income streams for often remote communities. This chapter explores the wider regulatory and policy context, the roots of community energy policy, the diversity of practice, the implications of increased devolution on community energy policy, the actual and potential benefits arising from community energy and the obstacles to increasing its share of the renewable energy market in Scotland.

Chapter 4 analyses how Scotland has attained a leading role in endeavours to stimulate and grow the nascent marine energy sector. Establishing technology leadership in the sector would result in substantial benefits for Scotland, as the world moves away from hydrocarbon dependency towards renewable energy sources. Scotland is almost uniquely placed to establish itself as the world-leading centre of marine energy technology, potentially resulting in major employment and global export opportunities for the supply of goods and services for wave and tidal schemes throughout the world. Although marine energy derived from wave and tidal energy offers a more predictable alternative to intermittent renewable energy technologies, such as wind and solar, and significant hurdles in the marine energy technology development cycle have already been crossed, the main challenge now is to prove commercial viability.

The remaining two chapters of Part One expand the analysis of low-carbon energy policy and practice in Scotland. Chapter 5 analyses renewable heat, highlighting that no single area of policy has a greater potential to undermine the Scottish Government's energy and climate change targets than the provision of renewable heat, and if it fails then those hit hardest will be the fuel-poor, and particularly those in the islands and rural areas. This chapter explores how factors including the poor condition of Scottish housing, high levels of fuel poverty, lack of investment in infrastructure and the persistence of a belief in the validity of the outputs of flawed models are leaving Scottish communities exposed to both the changing climate and changing energy prices beyond 2020. It concludes with an alternative vision of renewable heat in Scotland that could be enough to avert the worst of this perfect storm of events.

The particular role of nuclear power is investigated in Chap. 6. This chapter examines the role of nuclear energy on current Scottish energy policy, an underexplored area of value whether Scotland remains in the UK, secures independence or further devolution post-Brexit. A recurrent theme in the analysis is that whether one is for, against or indifferent to new nuclear energy development; it highlights a major gap in Scotland's energy and environmental policy goals. Too often, the Scottish Government perspective has been reduced to a low-carbon energy development debate between nuclear and renewables, with little reflection on how to reduce fossil fuel dependency. Aspirations to being a low-carbon economy, a global leader in climate change and to decarbonising its electricity market means Scotland needs to tackle the issue of how to stop burning fossil fuels.

Part Two focuses on the challenges ahead. Chapter 7 addresses the dual issue of energy efficiency and behavioural change in attempts to reduce energy demand. The UK's road to energy efficiency has been paved with mediocrity, and whilst Scotland has done somewhat better in terms of getting energy efficiency measures installed it has still failed to do much about the underlying problems of the poor condition of its housing stock and the need to change occupant behaviour. This chapter discusses how the frequently cited problem of the prevalence of 'hard-to-treat' properties is rarely a technical barrier, and questions why the Scottish Government has not done more to learn from the failings of both the UK and Scottish energy efficiency schemes. It also explores how more could be done to address the biggest problem for reducing energy demand, namely how to design policies that are sensitive to the fact that humans are innately human.

Chapter 8 describes the UK renewable energy policy reforms implemented in 2015, placing these events in a historical and European context and analysing the discourse and underlying rationale from which they emerged. Presenting a detailed overview of reforms implemented around renewable energy support mechanisms, it examines the implications for different technologies and scales of deployment, focusing on renewable electricity. Drawing on reports and statements issued by the newly elected 2015 government and analysing reforms in the context of the UK's broader energy and climate mitigation policy, this chapter shows that historical and deep-rooted party political narratives around renewable energy have remained virtually unchanged since debates leading up to the 1990 Electricity Act. Finally, looking forward, the prospects for Scottish Government public support and for 'subsidy-free renewable energy' are explored.

The remaining two chapters of this part of the book focus on the two main strands still resonating from the ongoing independence debate: further devolution and possible independence. Although Scotland voted in the 2014 independence referendum to remain in the UK, Chap. 9 notes that this offered a window of opportunity for a new devolutionary settlement, with implications for renewables given their importance to the Scottish Government. This chapter seeks to answer whether the new settlement results in a more cohesive set of devolved powers in the key area of Scottish renewable electricity policy, by assessing the Scotland Act 2016 as recommended by the Smith Report alongside both the Scottish Government's aspirations pre- and post-referendum and the implications for renewable electricity technology deployment regarding the 2020 target and beyond. It is argued here that further devolution did not alter the status quo. Furthermore, the nature and scope of the new powers reaffirm the view that Scotland should remain merely a consultative party in the governance and management of the UK renewables policy. Analysis of the new powers shows that the level of influence held by the Scottish Government is insufficient to effect real change. This leads to the conclusion that the new devolved powers act as a constraint to the realisation of Scottish-specific renewable electricity policy, with potential impact on large-scale deployment going forward. This has significant implications given calls for another referendum following Brexit.

The final chapter assesses the potential impact of independence on Scottish electricity in the case of independence. It does seem likely that it would be in British as well as Scottish interests to maintain the British electricity system much as it is managed now, although the Scottish Government would have to shoulder financial responsibility for new renewable energy deployed after independence. This could be relatively cheaper as time moves on, although long-term contracts for reasonable prices to be paid for renewable energy would still have to be issued by the Scottish Government and costs borne by solely Scottish consumers if renewable energy can be expected to grow. Certainly, a Scottish Government could support substantial development in onshore wind without great increases in electricity prices, and thus reverse the effective ban on funding future onshore wind imposed by Westminster. However, regardless of this, it seems unlikely that renewable energy will yield the tax receipts that have been supplied by oil extraction. A trend towards electric cars is likely to restrain oil prices and thus oil tax revenues in the future.

1.3 What Next?

As we go to press, Scotland's future still hangs in the balance. Whilst First Minister Nicola Sturgeon and the SNP remain in power, the Scottish Government will remain committed to opposing withdrawal from the European Common Market and resisting restrictions on the freedom of movement of labour that are critical to Scotland's strong and growing renewable energy sector. However, despite the SNP now constituting the third-largest party at Westminster, and arguably the only effective opposition to the ruling Conservative Party, it is clear that Scotland's interests are being given short shrift as the UK moves towards a seemingly inevitable hard exit from the European Union.

Should this come to pass, it will mean the Scottish Government, and the people of Scotland, will have to decide whether it will be better for the country to remain in the UK and make the best of whatever settlement is finally agreed, or vote for a new future within Europe but outside the UK. The outcome of that decision will have a huge impact on Scotland's energy future and, as these contributions to the debate show, declaring independence would carry with it the sole responsibility for the country's future development of renewable energy. A bright future is certainly possible, and easier to realise as an independent member of the European Union, however, to face it uncritically would be ignoring the evidence that not all the Scottish Government's failures to date can simply be blamed on Westminster, and that the hardest challenges may yet lie ahead.

Aye? Naw? Mibbe?

Notes

1. Aye—Yes in answer to an affirmative or negative question. Naw—No, the negative reply to a question, the word used to indicate denial, disagreement, refusal or contradiction. Mibbe—Perhaps, possibly.
2. See Wood (2017) for a debate on whether renewable energy has reached a tipping point at https://sputniknews.com/radio_level_talk/201701121049520046-renewable-energy-tipping-point/.

Authors' Biography

Dr. Geoffrey Wood is a Teaching Fellow in International Energy Law and Policy at Stirling University. An energy and environmental law, policy and regulatory specialist, his work focuses on the development of legal and governance frameworks for sustainable energy transitions to optimise policy delivery in terms of environmental, social, economic, political and technological outcomes. He has published over 30 articles, books, chapters and other publications. Currently also working as a Researcher and Contributing Lecturer at Dundee University, Geoffrey has previously held research, lecturing and consultancy positions for various organisations, including the Scottish Government and the Offshore Renewables Institute.

Dr. Keith Baker is a Researcher at the School of Engineering and the Built Environment, Glasgow Caledonian University and a co-founder of the Initiative for Carbon Accounting (ICARB). He specialises in energy in the built environment, with a particular focus on fuel poverty, energy efficiency and sustainable building materials.

CHAPTER 2

Large-Scale Renewables: Policy and Practice Under Devolution

Geoffrey Wood

2.1 Introduction

The 2014 referendum on Scottish independence has rightly crystallised attention on the renewable electricity sector. Increasingly portrayed as one of the success stories of renewable electricity, both within the UK and abroad, Scotland is committed to a dramatic increase in the level of renewable electricity technology (RET) deployment within a very tight timetable to meet one of the most ambitious electricity generated from renewable sources (RES-E) targets in the world: 100% equivalent of gross electricity consumption from renewable energy sources by 2020 (Scottish Government 2011). This equates to the need to deploy 8 GW in 5 years, from around 8 GW of capacity in 2016. By all accounts a demanding target, so far all previous targets have been met on time or surpassed, including the 2011 target of 31% which was exceeded by 5% and the 2015 interim target of 50% (Scottish Government 2016). In stark contrast, the UK was 2 years late in achieving the 2010 RES-E target of 10% (Department for Energy and Climate Change [DECC] 2013a).[1]

Although the Scottish public voted no in the independence referendum, panic in the latter stage of the referendum by the three main

G. Wood (✉)
School of Law, University of Stirling, Stirling, UK
e-mail: geoffrey.wood@stir.ac.uk

G. Wood and K. Baker (eds.), *A Critical Review of Scottish Renewable and Low Carbon Energy Policy*, Energy, Climate and the Environment,
DOI 10.1007/978-3-319-56898-0_2

pro-union political parties (Labour, Conservatives and Liberal Democrats) resulted in the offer of substantial new devolved powers to the Scottish Government. Set up to oversee the process to take forward devolution commitments on further powers, the Smith Commission swiftly published its recommendations for further devolution (Smith Commission 2014). With a number of the recommendations already legislated for in the Scotland Act 2016, understanding the context and the implications of the existing devolutionary settlement for renewable deployment going forward is critical before looking at the new powers and the potential implications for Scotland gaining additional control over energy policy and related areas (see in particular Chap. 10). Therefore, this chapter will look at the devolutionary settlement as it stood immediately prior to the Smith Commission.

Devolution is an important consideration in leading towards both a separation and divergence of powers, policy and practice with the emergence of an indigenous and increasingly confident Scottish renewable electricity policy. Devolution has also had a particular impact on Scotland, due to the greater powers devolved to the Scottish Government in contrast to Wales.[2] Equally important, however, and connected to the scope and use of devolved powers is the approach to RET deployment by the Scottish Government albeit with the caveat that overall energy policy remains a reserved matter to the UK Government in Westminster.

This chapter examines the development of Scottish renewable electricity policy under devolution and the implications of devolution for the deployment of large-scale RETs. As such, this chapter focusses on the divergence in policy and practice in the Scottish Government's approach to promoting large-scale RET deployment under devolution. This is all the more relevant given that the renewables sector is once again entering a new phase of radical reform with changes to the fundamental way in which large-scale renewable electricity technologies are promoted via the ongoing Electricity Market Reform (EMR) process. Introduced in April 2014, the Contracts for Difference Feed-in Tariff (CfD FiT) has already replaced the Renewables Obligation (RO) mechanism one year earlier than scheduled.[3] Furthermore, this chapter aims to add to the debate by clarifying the context in which decisions on a future Scottish renewable electricity policy must be based.

2.2 Renewable Electricity Deployment in Scotland

In the last 13 years, RET deployment has almost quadrupled to almost 8 GW of installed capacity and electricity generated from renewables accounts for 57.4% of Scotland's gross electricity consumption in 2015, overtaking all other power sources in terms of output including nuclear power (Scottish Government 2016). In a country with just 8% of the UK's population and 32% of its landmass, Scotland also accounts for around a quarter of UK capacity and 70% of RET deployment in the devolved administrations.

RET deployment capacity has increased year by year since the introduction of the RO in 2002. The main success story of Scottish RET deployment to date is onshore wind power. This one technology accounts for over two-thirds of total RET capacity in Scotland and 60% of total UK installed capacity (DECC 2016). Excluding hydro power, a legacy of the nationalised construction of large-scale reservoir dams after World War II, onshore wind accounts for almost 90% of all capacity, and almost 90% of average annual new-build for the period 2002–2015 in Scotland was for onshore wind farms. Despite the addition of only one new major hydro plant in the last 50 years, hydro power still accounts for a quarter of total installed capacity. By 2015, both onshore wind and hydro power also dominated RES-E generation, accounting for 90% of total generation output.

In stark contrast, the other technologies have shown limited deployment. This can be partly justified by reasons of technological maturity and resource availability. Onshore wind is one of the most mature and cheapest RETs with over two decades of deployment experience in the UK. Scotland has significant onshore wind resources, and technology development has increased over the years to utilise this through larger and more efficient turbine designs and associated increases in tower height. It is an obvious choice for government support and uptake by market participants. The opposite is true for marine renewables including wave and tidal power, despite Scotland having 10 and 25% of Europe's potential wave and tidal reserves, respectively. This is because they represent immature technologies primarily in the R&D or demonstration stage. For solar photovoltaic (PV), Scotland has significantly less solar radiation levels than the rest of the UK, and this technology is primarily subsidised through the small-scale feed-in tariff governed at the UK level by DECC. There is also only 358 MW (5% of total Scottish capacity) of all biomass and waste technologies. Such limited deployment has been a consistent trend over the last 10 years and more and this is particularly the case in comparison with the UK overall, with around 5.2 GW of installed capacity. There are a number of reasons for this, including sustainability issues (CO_2 emissions and sourcing of biomass fuel-stock particularly from abroad) and the importance of biomass towards meeting the renewable heat sectoral target in addition to other parts of the economy.

With 25% of Europe's resource potential, offshore wind represents a key technology for the Scottish Government's strategic sustainable economic vision, with manufacturing, supply chain and job creation opportunities. It also has the potential to deploy at significant scale towards meeting the 2020 RES-E target and is perceived to avoid a number of barriers that have constrained onshore renewable deployment (particularly onshore wind) including planning, land use and public opposition. In contrast to onshore wind, though, there has been very little offshore wind deployment to date. Scotland has three operational offshore wind farms with an installed capacity of 197 MW, less than 5% of the total UK capacity for this technology of 5.1 GW with the majority located in English waters. However, again there are a number of reasons for this. Only 180 MW of capacity was located in Scottish waters out of a total of around 9 GW under the early Crown estate (CE) offshore wind leasing rounds (1 and 2). Subsequent leasing agreements did involve a higher proportion of Scottish sites but this occurred much later: Scottish Territorial Waters Round (2009) and Round 3 (2010). Importantly, the later rounds are planned in deeper waters farther from shore which increases the complexity, cost and time to develop the proposed projects.

2.3 What Has Devolution Brought to the Renewables Table?

Following a referendum on Scottish devolution held on 15 September 1997 by the then recently elected Labour Government, with 74% voting in favour of a Scottish Parliament, the legislative framework for Scottish devolution was set out in the Scotland Act 1998. Wales and Northern Ireland also voted in favour of devolution, although on different terms from the Scottish referendum (Ross 2012). Although the passage of the Act represented significant constitutional reform for the devolved administrations and the UK by kick-starting the devolution process, the separate nations have always retained their respective identities. Since the Acts of Union united the Kingdoms of Scotland, England and Wales to form the Kingdom of Great Britain (GB) over 300 years ago in 1707, Scotland has always maintained its own distinctive identity, legal and education systems and other aspects of civic life. What devolution has in effect brought about, in the last decade and a half, is political decision-making on key issues to the respective nations at a lower tier of governance, although the degree of devolved powers varies between the administrations.

Under devolution, Scotland now has a Scottish Parliament and a Scottish Government (originally an Executive) sitting at Holyrood in Edinburgh. The legislative powers for energy and related areas are separated into reserved matters remaining under the full jurisdiction of the UK Government with all other matters not listed in the Act deemed to be devolved to the Scottish Government. As expected, constitutional and fiscal matters are reserved at the UK level. As mentioned previously, energy policy is also a reserved matter although, as will be argued below, control of centralised policy making is not so clear cut: rather than being set in stone (or more accurately in the Scotland Act 1998), the situation is somewhat more fluid. In general, however, this means that the generation, transmission, distribution and supply of electricity, the ownership of, exploration and exploitation of oil and gas deposits, coal (including its ownership and exploitation) and nuclear energy and nuclear installations are reserved to the UK Government (Paterson 2013). Devolved matters include the environment, planning and economic development.

Dividing responsibilities between the UK and Scottish Government in this way, however, does not lead to a clear separation of powers in practice. There are many factors that need to be considered in attempting to meet renewable energy policy objectives, particularly in the case where two countries have differing policy considerations. These include economic, technical, social, environmental and behavioural issues, often influenced by events both within and outside the UK. Energy policy, then, sits on the dividing line of powers and legally binding and non-binding obligations and targets. What needs to be remembered is that it is the UK, as the sovereign state, that holds key responsibilities for meeting targets: the EU 2020 and 2030 targets for renewable energy and climate change. As such, it is the UK Government, primarily through the DECC and the Treasury, that designs the wider electricity market and the main subsidy mechanisms to promote renewable electricity, including the current RO to financially incentivise large-scale RES-E generation and the replacement CfD FiT mechanism via the ongoing EMR process. They also hold oversight responsibility for regulating both the energy sector and energy networks via the Office for Gas and Electricity Markets (OFGEM), a pan-UK independent energy regulatory body alongside other bodies including the CE and National Grid. It is also the UK Government that participates directly in negotiations at the international level on the direction of current and future energy relevant policy.

2.4 What Does This Mean for Scotland?

Devolution has resulted in the Scottish Government gaining a number of levers of control over the evolution of the future electricity generation mix and in addressing barriers to deployment. Often perceived as key barriers to increasing RET capacity, these include planning, renewable electricity incentives and the transmission network.

2.4.1 Planning

By transferring control over the onshore and offshore planning system, devolution has resulted in the Scottish Government gaining substantial control over realising its renewable policy objectives. The devolution of planning permits the Scottish Government to ultimately decide which types of power generation can take place within Scotland's territorial jurisdiction: coal plant (yes with strict caveats); gas and other thermal generation including biomass (yes with caveats); non-thermal renewables such as wind power (an unqualified yes) and new nuclear power (a definite no). With regard to major energy infrastructure, devolution has transferred powers to issue planning consent for onshore power stations with an installed capacity of 50 MW or above and power lines with a nominal voltage exceeding 20 kV or more from Westminster to Scottish Ministers. Onshore power stations and lines below these thresholds fall under the remit of the relevant local planning authority and the Town and Country (Scotland) Act 1997 applies. In relation to the marine environment, the Marine (Scotland) Act 2010 legislates for marine planning and licensing and conservation activities in Scottish inshore regions (0–12 nautical miles, or nm). The UK Marine and Coastal

Access Act 2009 executively devolved marine planning and licensing and conservation powers in the offshore region (12–200 nm) to Scottish Ministers.[4]

The devolved control of planning has enabled the Scottish Government to actively support certain RETs and mitigate planning problems to a greater extent than realised in other parts of the UK. This has been an important factor in the consistent growth of onshore wind capacity. Arguably, the most contentious renewable technology, primarily due to landscape and land use concerns (Nadaï and van der Horst 2010), RET deployment in Scotland is also dominated and currently dependent on this one technology. In Scotland, as with other parts of the UK, there is also growing opposition to the technology in the planning system (Warren and McFadyen 2010). This is not surprising. By 2012, there were already 160 operational wind farms in Scotland with another 152 under or awaiting construction and a further 235 pending a planning decision (Wood 2013). With a significant proportion of deployment required to meet the 2020 target anticipated to come from onshore wind, this technology has become a very emotive and politicised issue.

Acknowledging the increasing pressure of onshore wind on both the planning system and public opposition and the challenging 2020 target, the Scottish Government has used devolved planning powers to centralise control over the consenting process for a number of different types and scales of developments (Wood 2010). Whilst approval rates for wind farms that fall under the jurisdiction of Scottish Ministers (>50 MW installed capacity) averaged 87% over the period 2007–2012, approval rates for local planning authority consented projects (<50 MW) fell from 75 to 50% during the same period (Wood 2013). In contrast, although approval rates in England under the remit of the Secretary of State averaged 92%, local planning authority consented projects declined from 72 to 29% over the same period. Furthermore, the higher approval rate for >50 MW projects conceals the fact that England has significantly less operational onshore wind capacity than Scotland, around 40% (DECC 2016).

The Planning Etc. (Scotland) Act 2006 introduced a hierarchy of planning consisting of national (projects of long-term national significance), major (including generating plant with an installed capacity >20 MW), local (<20 MW capacity) and minor (permitted or given deemed planning permission) developments.[5] Scottish Ministers have potentially significant influence over any projects that fall within the first three levels: the power to designate national developments through the National Planning Framework, the ability to call-in any national or major projects to speed up decisions, and direct any local developments to be dealt with as if it was a major development (Wood 2010). Scottish Ministers also play a role in the appeal process for major and local projects. Over the period from May 2007 to December 2014, 39% of wind turbine-related appeals referred to the government after an application was refused by a local planning authority were allowed (Scottish Government 2014). In contrast, the corresponding situation in England has become increasingly politicised with the Secretary of State for Communities and Local Government removing decision-making responsibility from local planning authorities. Out of 50 recovered projects which were at the appeal stage, only 10% have been allowed out of 19 projects where decisions have been reached with 5 projects previously recommended for approval by the Planning Inspectorate (RenewableUK 2014a). At the local planning authority level, the Scottish Government has also used spatial planning to designate areas specifically for onshore wind deployment, reaffirming the importance of the technology. The *Scottish Planning Policy* document requires planning authorities to determine suitable areas for >20 MW onshore wind farms (and to consider <20 MW projects) in development plans (Scottish Government 2010).[6]

Control of marine planning and licensing has granted the Scottish Government more effective powers for offshore RETs, enabling the creation of a one-stop shop for offshore wind, wave and tidal developers to obtain planning consent and relevant licences required to develop generating plants in Scottish waters (Scottish Government 2012a). Resulting in a more joined-up process that promotes close working relationships between developers and consulting bodies, this has simplified and streamlined the process for developers and regulators in comparison with the rest of the UK. Furthermore, the executive devolution of marine planning has enabled the Scottish Government to centralise control to a higher degree than that of the onshore planning regime. There are two main reasons. In line with the Electricity Act 1989,[7] virtually all offshore RETs will fall under the remit of Scottish Ministers. Sub-1 MW projects fall under the remit of the new statutory strategic regulator for marine-related functions in the relevant waters, Marine Scotland, a Directorate of the Scottish Government (Scottish Government 2015a). Essentially, Scottish Ministers will retain control over marine planning and licensing for all commercial-scale developments and initial small-scale projects, with particular importance for early-stage marine technologies. However, the Scottish Government has no devolved powers over the granting of leases for offshore RETs, including half the Scottish foreshore and virtually all territorial waters out to 200 nm. This is governed by the CE, a UK-wide property portfolio owned by the Crown and governed by an Act of the UK Parliament (Crown Estate Act 1961) (Crown Estate 2014). As such, the CE plays a major role in the development of the Scottish offshore wind, wave and tidal stream energy industry although it is not involved in the planning and generation licensing process.

2.4.2 Renewable Electricity Incentives

Under devolution, the Scottish Government also acquired a degree of operational control over the ROS subsidy mechanism.[8] In practical terms, this primarily meant the ability to set subsidy levels for individual RETs that differed from those in the rest of the UK and changes to criteria determining the eligibility of RETs to receive subsidy via the mechanism. As a tool for supporting technologies, these powers have been used in innovative ways to great effect in Scotland to maintain investor confidence and policy stability for developers. Marine renewables have particularly benefitted from this approach. The provision of consistently higher subsidies for wave and tidal power technologies in Scotland under the ROS than was available elsewhere in the UK enabled the Scottish Government to overcome a somewhat *laissez-faire* attitude that characterised the UK Government's approach until recently.[9]

Indeed, the Scottish Government has been very proactive in both policy development and policy in practice to a greater extent than most countries engaged in this evolving sector. In recognition that marine renewables are typically at the prototype or demonstration stage (pre-commercial) and largely brought forward by small sized companies, the Scottish Government has also funded initiatives from discretionary government spending to bridge the gap between research, design and development on the one hand and deployment and commercial operation on the other (Scottish Government 2015b). The Scottish Government also showed foresight in supporting the European Marine Energy Centre (EMEC) to bring forward device testing in real marine conditions and the £10 million Saltire Prize to drive innovation in the sector. With the global race to de-risk and commercially deploy these technologies, there are substantial economic benefits, in terms of both developing domestic and export markets, to be realised from leading technological development and developing a viable marine renewables sector in comparison with the offshore wind (Wood 2010). Scotland has already positioned itself as a world leader, and this has enormous political benefit for the Scottish Government.

Another example of policy divergence between the Scottish and UK Government approaches is the new ROS only offshore wind technology bands offering increased subsidies for floating or innovative turbines and demonstration turbines. Proving these technologies would allow the more optimal utilisation of Scotland's offshore wind potential, the bulk of it in very deep waters far from shore (RenewableUK 2014b). Although DECC has ruled out separate subsidy support for these emergent technologies (DECC 2012a), this further highlights the capacity for policy innovation and experimentation at the sub-national level. Wave, tidal power and innovative or demonstration stage offshore wind turbines, however, are long-term technology options as evidenced by the very limited deployment to date. The Scottish Government has also used its devolved powers to both promote more stringent environmental objectives and maintain policy stability for those technologies it views as key to increasing capacity in the near-term. With regard to the former, the Scottish Government has made clear its preference for biomass to be utilised for heat or combined heat and power (CHP) generation by setting stricter eligibility criteria to include sustainability issues (CO_2 emissions and sourcing of biomass fuels particularly from abroad) and changes to the subsidy offered under the ROS for certain biomass technologies to limit greenhouse gas (GHG) emissions, meet non-power renewable targets and protect other key industries (Scottish Government 2012b). Regarding the latter, when the UK Government immediately launched an unscheduled banding review seeking further cuts to onshore wind after having already reduced subsidy levels by 10% in 2012, the Scottish Government acted decisively to rule out further cuts and guarantee support until 2017 (Pinsent Masons 2012). Although the UK Government ultimately decided not to impose additional cuts, the proposal alone created uncertainty with the decision pending for over a year (DECC 2013b). This is important. Without short-term certainty and longer-term visibility that render financial and political risks reasonably predictable and manageable, projects will not be as viable or attractive and this could impact on deployment (Plant 2013). With recent studies indicating between 675 and 1200 MW of new hydro potential in

Scotland, the Scottish Government also diverged from the UK position by ruling out cuts to hydro power in contrast to a cut of around a third of subsidy under the RO (Pinsent Masons 2012).

However, two recent developments at the UK level have stripped the Scottish Government of its powers over the operation of renewable electricity mechanisms. Firstly, section 55 of the Energy Act 2013 contained provisions for the Secretary of State to close the RO from 31 March 2017, enforceable from the Act entering into law.[10] The point here is not that the mechanism would be closed or at that specific date, this was known since the early stages of the EMR process, but rather that the UK Government, without any prior consultation or discussion with the Scottish Government, removed Scottish Ministers and the Scottish Parliament of powers and discretion already granted under devolution. Anyway, this point became moot when the UK Government, without warning, closed the RO one year earlier than scheduled. This action also had the practical effect of undermining the introduction of the new bands for floating and demonstration turbines, with the Scottish Government having to seek clarification from Westminster in order to assuage investor concerns (Scottish Government 2013). Secondly, the CfD FiT is a more centralised mechanism to financially support RETs than the RO (Energy and Climate Change Committee 2012). By design, the Scottish Government has none of the control over the new mechanism that it used to have under the ROS: it cannot include or exclude technologies, set the subsidy level (or strike price under the CfD FiT) and has no power over the process of contract allocation for new projects.

The significance of this cannot be over-emphasised. It removes virtually all control over the renewable electricity subsidy mechanisms. It is also important to recognise that despite well-documented concerns with the RO, specifically due to the type, design and operation of the mechanism (Wood and Dow 2011), the Scottish Government has never had the devolved competence or influence to address these fundamental issues. The main point here is that despite devolution, the Scottish Government cannot replace or fundamentally change the design of the mechanism. That is the prerogative of the UK Government. Furthermore, there is to be no socialisation of costs under the CfD FiT with regard to the setting of different subsidy levels for RETs in the devolved administrations (DECC 2012b). Where energy policy has been fully devolved in the case of Northern Ireland, the Northern Ireland Executive (NIE) has adopted the new mechanism in full to be administered on a UK-wide basis (Department of Enterprise, Trade and Investment 2014). In addition, although the NIE does have the ability to set different strike prices to reflect different market arrangements, it has agreed to the GB-wide strike prices (Northern Ireland Executive 2013). The alternative would be for additional costs to be met only by consumers in Northern Ireland and not across the UK as was the case under the RO. Yet the socialisation of costs was a key debate and potential stumbling block in discussions of Scottish independence. As with Scotland, the end result is the same, with innovative approaches to supporting RETs by the devolved administrations being effectively ruled out.

This also leads to the issue of how to support expensive technology options requiring long-term financial and policy support including offshore wind and marine RETs when they reach commercial-scale deployment. In particular, there are a number of challenges of a technical, economic, social and environmental nature that face offshore wind not just in Scotland but in the UK and abroad, including policy risk (Wood 2010). Both the UK and Scottish Governments have agreed to the need to reduce technology costs by a third by 2020 (Offshore Wind Cost Reduction Task Force [OWCRTF] 2012). However, contrary to the deployment experience from earlier CE rounds, costs have escalated from the mid-2000s and reductions are expected to occur only gradually to the mid-2020s (OWCRTF 2012). The downward trend in costs will only be achieved if supply chain constraints are addressed alongside the technology, construction, regulatory and financial de-risking of offshore wind through research and development and demonstration as deployment moves into deeper waters further from shore. This will require a concerted and sustained effort by all stakeholders involved in the sector: the UK Government, the devolved administrations and other countries within the EU and beyond, regulatory and other statutory bodies, developers (typically multinational and often state-owned to some extent) and supply chain companies, non-statutory organisations and the public who ultimately will pay for sector development through their energy bills (OWCRTF 2012). Critically, not all of the barriers to deployment lie within the Scottish Government's jurisdictional control.

However, policy risk at the UK level is threatening to derail offshore wind deployment in Scotland, with the sector recently hit by a number of setbacks for proposed projects across the UK. This is in addition to capacity attrition of other projects due to various reasons including public objections, technical and environmental concerns. Although not all of the cancellations are due to political or policy risk, key players in the sector have either cancelled or halted commitments to develop projects post-planning consent being obtained. This decision appears to have been taken mainly in response to the considerable political debate between the UK Government and the major energy companies over energy prices rises and the impact on the affordability of customer bills and uncertainty due to the EMR and lack of a post-2020 target.[11] The implications for the Scottish offshore wind sector are profound, with at best the delay and at worst the loss of around 3.6 GW of directly proposed capacity and billions in investment. If the total capacity offered for the Firth of Forth Round 3 zone is taken into consideration, this figure increases to almost 5 GW or a third of the 2020 target. Currently, only the Beatrice offshore wind project has been awarded a CfD FiT agreement, representing only around 600 MW. This has also negatively impacted on the Scottish Government's plans to more than double biomass power capacity in Scotland, with over 400 MW cancelled, despite half the capacity already receiving planning consent from the minister, essentially due to the same reasons stated by developers for the proposed offshore wind farms (BBC 2014). This also gives a strong indication of the highly political nature of energy policy in the UK.

2.4.3 Electricity Network

The transmission and distribution network is also considered a key barrier to deployment, with an unprecedented amount of grid capacity required to connect new renewables (Electricity Networks Strategy Group [ENSG] 2009). Grid problems will particularly affect onshore wind farms but increasingly offshore wind and future marine renewables as they continue to be deployed at scale. This means mostly onshore wind farms but with implications for offshore wind and future marine renewables. However, with the exception of planning, the Scottish Government has very little power over either the onshore transmission or distribution networks. It has no regulatory powers to allocate new upgrades and extension of the network, or change access rules to the grid or the charging regime. As the pan-UK energy regulator of the single GB electricity system, this is the remit of OFGEM with an important role for National Grid as the system operator (OFGEM 2014). Furthermore, on the policy and legislative side, it is DECC that introduces grid reforms and not the Scottish Government. As with the payment of subsidies for renewable energy, one of the key benefits of this approach is that the costs of building and maintaining the networks are socialised across GB. This is an important point given that the majority of the work going forward is to be located within the boundaries of Scotland (ENSG 2009).

Although OFGEM and DECC have been proactive in increasing network capacity, particularly in Scotland which is currently heavily congested, and implementing reforms such as the connect and manage regime to speed up connection times, the protracted debate on locational charging between the Scottish and UK Governments highlights this issue (The Guardian 2013). Locational charging, reflecting the cost of transporting power, imposes higher costs on Scottish generators compared to generators in the south of England due to being located farther from the area of greatest demand in the south of England; some generators receive subsidy due to being located in southern England. Although the Scottish Government has not formally been able to amend this, as part of the UK it has been able to influence thinking and OFGEM announced a change to the charging methodology in August 2013.[12] Another example of the disjointed devolution of powers that further highlight the arbitrary nature of energy devolution is the differences in strategic planning over the onshore and offshore electricity networks. As McHarg (2014: 1) states, '*Why should* [the Scottish Government] *be able to plan the development of offshore electricity networks, but have no equivalent powers over onshore networks?*'

In practical terms, then, despite overall energy policy being reserved to Westminster, substantial areas of energy policy have been devolved. The extent of existing devolved powers to Scotland, however, is largely piecemeal, and there is no guarantee that these powers will not be removed. Devolution has therefore not led to a black-and-white repertoire of powers. The Scottish Government does have the potential to exert influence over energy and renewable deployment at the Scottish-specific level. The crucial question is how much influence does the Scottish Government possess in the sphere of renewable technology deployment? Just as important, how is that power used? In a very real sense, devolution has both provided and legitimised the '*space*', whereby Scotland and the other devolved nations now at least have the potential to create their own energy policy, albeit constrained by the boundaries of devolution. Importantly, as these boundaries are not set in stone, the devolved administrations have the opportunity to engage with policy implementation and processes in Westminster through intergovernmental bargaining and negotiation at the formal (consultations, setting targets and producing policy documents) and informal (dialogue, behind the scenes agreements) level (Cowell et al. 2013). Devolution has also allowed the devolved administrations to set out their own distinctive policy strategies and priorities on the issue of renewable energy.

2.5 Different Visions and Divergent Approaches in Scotland and the UK

There is no doubt that formal and informal devolved powers are important. However, recent research investigating the impact of devolution on the promotion of renewable energy in the UK has shown that it cannot fully explain the different levels of success in increasing deployment: '*… simply possessing "powers" in the narrow legal or administrative sense may be of limited relevance without a disposition, capacity or will to deploy them in an effective manner for renewable energy. In short, "powers" is an insufficient explanation*' (Cowell et al. 2013: 2). Political support can be just as important, and there are reasons why this is particularly the case for renewable electricity.

In general, RETs are relatively expensive technologies to deploy (in terms of capital and operational expenditure), and they face a number of barriers to deployment fairly unique to this technology category: some technologies, like wind power or large-scale hydro, can have significant impacts on landscape and land use, whilst biomass can cause particulate pollution and result in unsustainable forestry practice. Other technologies are regarded as immature with limited deployment experience, including wave and tidal power, and offshore wind. Because of the novel characteristics of these technology options, in addition to their small capacity factors requiring relatively large individual plant sizes (in terms of square metres), resulting in the need for more developments than conventional power sources like fossil fuels and nuclear power, political support is arguably a critical prerequisite for the promotion of renewable energy. In other words, a stable and coherent political strategic vision is required to overcome a number of challenges given the current need for financial, policy, legislative and regulatory support for the majority of such technologies.

Upon winning the 2007 Scottish elections, the SNP immediately set renewable energy as one of its core priorities and objectives in delivering ambitions for a greener Scotland in order to achieve sustainable economic growth (Scottish Government 2007). Based on the substantial potential of Scotland's onshore and offshore renewable reserves, the economic strategy of the Scottish Government was to become the 'Saudi

Arabia' of renewable energy with the potential to deploy up to 60 GW of renewable electricity capacity, more than 10 times current peak demand (Business Green 2008). The distinctive Scottish emphasis on renewable energy has been consistently reiterated and reinforced through a cohesive and stable vision going forward. This vision has been backed-up and developed by various policy documents including the *Electricity Generation Policy Statements* and the *2020 Routemap for Renewable Energy in Scotland* alongside additional Scottish-specific initiatives to promote renewable deployment.

In contrast, the previous UK Coalition Government (the Conservatives and Liberal Democrats) vacillated between support for nuclear power, shale gas, carbon capture and storage and renewables, and the election of a majority Conservative Government in 2015 looks unlikely to change this: indeed, they appear more stable at least in terms of showing less confliction and desire for supporting renewables. Nuclear power and shale gas benefit the most from political support, as evidenced by the tortuous EMR process over the last 5 years which seems to be an attempt to underpin new nuclear build whilst avoiding the appearance of subsidising it and the almost gung-ho push for a rapid expansion of shale gas extraction based on US success in exploiting its domestic resources. It is clear that political motivation to support renewables falls far short of that on offer for other 'chosen' technologies. The bitter rift between the two Coalition parties over the future of onshore wind, with the Conservative Party's proposal to cap the future capacity of the technology, is one such example. From the laudable but utterly vague slogan of becoming the 'Greenest Government Ever', in the space of just 4 years the Coalition increasingly moved towards supporting nuclear power and fossil fuels through strong policy commitments, financial incentives and addressing regulatory barriers to their deployment.

2.6 Conclusion

Two key points can be made from reviewing the existing devolutionary settlement regarding RET deployment in Scotland. First, devolution has resulted in significant benefits for the Scottish Government in realising its renewable energy ambitions, in terms of policy and practice. This can be seen in the approach to making full use of planning functions and tailoring the ROS subsidy mechanism to promote those technologies seen as particularly important to Scottish ambitions, including onshore and offshore wind and marine technologies. Such initiatives have also been driven by consistent and stable policy aims and objectives as Scotland seeks to carve out a distinctive Scottish-specific energy policy. Even in areas where control lies firmly within the jurisdiction of pan-UK institutions, for example in decision-making over where and when to upgrade the electricity network, the Scottish Government has been particularly vocal and determined, although network enhancement is also of advantage to the UK as well.

Second, it is also clear that devolution has not resulted in a clear demarcation of powers between Westminster and Holyrood. This is to be expected due to the complex and systemic nature of energy issues, the different policies and strategic aims that exist at the sub-national level and the national level, and the fact that Scotland remains a part of the UK. However, despite the ongoing process of devolution (prior to the Smith Commission) resulting in Scotland gaining legislative competence and the legitimised capacity to influence UK energy policy from a Scottish perspective, this has produced an existing devolutionary settlement for renewable energy and indeed wider energy issues that is largely individualistic, piecemeal and arbitrary in terms of what is reserved to the UK Government and what is devolved to the Scottish Government. Of further concern is the removal of existing devolved powers by Westminster and the lack of a guarantee that any of the remaining powers will not be clawed-back at some future date, particularly with the introduction of the CfD FiT mechanism where virtually all control already lies in the hands of Westminster. Surprisingly, there does not appear to have been any real discussion of the appropriate balance between devolved and reserved powers with regard to what would be optimal in terms of policy delivery of RETs. The reason why this is surprising is obvious, given the important contribution of Scottish-based RET deployment to domestic and international renewable and climate change targets, energy security, economic and employment issues at the devolved and overall UK level. Surely, a comprehensive and cohesive set of devolved powers over renewable electricity would be advantageous not only to Scotland but also to the UK overall. At the very least, it is an issue that should be investigated further.

Notes

1. It should be noted that all the chapter contributions were completed with final corrections in June 2016 prior to the new Conservative government coming to power with Theresa May as Prime Minister and the now new Department of Business, Energy and Industrial Strategy (BEIS).
2. Energy policy is already fully devolved to Northern Ireland.
3. At the overall UK level, the current subsidy mechanism is collectively called the Renewables Obligation (RO). In practice, they refer to three complementary obligations with different legal basis and variations in subsidy levels and eligibility criteria. These are the Renewables Obligation Scotland (ROS, Scotland), Renewables Obligation (RO, England and Wales) and the Northern Ireland Renewables Obligation (NIRO, Northern Ireland).
4. Onshore and offshore developers are required to apply for section 36 (power station) or section 37 (power line) consent from Scottish Ministers under the Electricity Act 1989 to construct, extend or operate a generating plant. Onshore developers must also apply for planning consent; in contrast, separate planning permission is not required to be obtained by applicants for offshore generators as section 36 consents and marine planning and licensing are considered together. Furthermore, developments with a capacity of 1 MW or less are exempt from section 36 requirements.
5. The Planning Etc. (Scotland) Act section 5.
6. This is important as the development plan is the basis of decision-making in the planning system; effectively what is not included in the development plan should not be granted consent (see Scottish Government 2010).
7. Electricity Act 1989 section 36.
8. The Renewables Obligation (Scotland) Order 2002, SSI 2002/163.
9. Differentiated support was provided even prior to the introduction of technology banding under the RO, through the Scottish-only Marine Supply Obligation which ran from 2007 to 2009. (cf. Wood 2010).
10. The Renewables Obligation Closure Order 2014, SI 2014/2388.
11. Key companies include Scottish and Southern Energy, Scottish Power, Centrica, DONG Energy, E.ON, RWE Innogy, Statoil, Statkraft and Masdar (cf. Scottish and Southern Energy 2014).
12. Although the new measures have not removed locational charging, the cost of transmission charges will be lower for Scottish generators than under the previous methodology (cf. OFGEM 2013).

References

BBC. 2014. Forth Energy Pulls Out of Scottish Biomass Projects—27 March 2014 (online). http://www.bbc.co.uk/news/uk-scotland-26767705. Accessed 10 Dec 2014.

Business Green. 2008. Scotland Aiming to Become "Saudi Arabia of Renewable Marine Energy": "Sheik" Salmond to Offer £10m to First Team to Deliver Commercially Viable Marine Energy—3 Dec 2008 (online). http://www.businessgreen.com/bg/news/1804187/scotland-aiming-saudi-arabia-renewable-marine-energy. Accessed 22 Aug 2014.

Cowell, R., G. Ellis, F. Sherry-Brennan, P. Strachan, and D. Toke. 2013. Promoting Renewable Energy in the UK: What Difference Has Devolution Made—Initial Findings, 23rd January 2013 (online). http://cplan.subsite.cf.ac.uk/cplan/sites/default/files/DREUD-FullReport.pdf. Accessed 22 Sept 2014.

Crown Estate. 2014. Energy and Infrastructure (online). http://www.thecrownestate.co.uk/. Accessed 22 Sept 2014.

Department of Energy and Climate Change [DECC]. 2012a. Government Response to the Consultation on Proposals for the Banded Support Under the Renewables Obligation for the Period 2013–17 and the Renewables Obligation Order 2012—July 2012 (online). https://www.gov.uk/government/uploads/system/uploads/attachment_data/file/42852/5936-renewables-obligation-consultation-the-government.pdf. Accessed 22 July 2014.

Department of Energy and Climate Change [DECC]. 2012b. Annex D Institutional Framework: Delivering EMR—November 2012 (online). https://www.gov.uk/government/uploads/system/uploads/attachment_data/file/65638/7080-electricity-market-reform-annex-d.pdf. Accessed Dec 2014.

Department of Energy and Climate Change [DECC]. 2013a. Energy Trends—December 2013 (online). https://www.gov.uk/government/uploads/system/uploads/attachment_data/file/266403/et_dec_13.pdf. Accessed Dec 2014.

Department of Energy and Climate Change [DECC]. 2013b. Onshore Wind Call for Evidence: Government Response to Part A (Community Engagement and Benefits) and Part B (Costs)—June 2013 (online). https://www.gov.uk/government/uploads/system/uploads/attachment_data/file/205423/onshore_wind_call_for_evidence_response.pdf. Accessed 02 Jan 2015.

Department of Energy and Climate Change [DECC]. 2016. Energy Trends Section 6: Renewables—Renewable Electricity Capacity and Generation (ET 6.1) (online). https://www.gov.uk/government/statistics/energy-trends-section-6-renewables. Accessed July 2016.

Department of Enterprise, Trade and Investment. 2014. Electricity Market Reform—9 October 2014 (online). http://www.detini.gov.uk/electricity_market_reform. Accessed 15 Dec 2014.

Electricity Networks Strategy Group [ENSG]. 2009. ENSG 'Our Electricity Transmission Network: A Vision for 2020' Full Report (online). http://www.ensg.gov.uk/assets/ensg_transmission_pwg_full_report_final_issue_1.pdf.

Energy and Climate Change Committee. 2012. Draft Energy Bill: Pre-legislative Scrutiny (HC 2012-13, 275 I) (online). http://www.publications.parliament.uk/pa/cm201213/cmselect/cmenergy/275/275.pdf. Accessed 02 Dec 2014.

McHarg, A. 2014. Energy Policy Devolution and the Smith Commission: Blog for the Scottish Constitutional Futures Forum, 2 December 2014 (online). http://www.scottishconstitutionalfutures.org/OpinionandAnalysis/ViewBlogPost/tabid/1767/articleType/ArticleView/articleId/4766/Aileen-McHarg-Energy-Policy-Devolution-and-the-Smith-Commission.aspx. Accessed 05 Dec 2014.

Nadaï and van der Horst. 2010. Editorial: Wind Power Planning, Landscapes and Publics. *Land Use Policy* 27: 181–183.

Northern Ireland Executive. 2013. Foster Announces Northern Ireland Support Levels for Large Scale Renewable Electricity From 2016: 20 December 2013 (online). http://www.northernireland.gov.uk/news-deti-201213-foster-announces-northern. Accessed 15 Dec 2014.

Office for Gas and Electricity Markets [OFGEM]. 2013. Ofgem Recommends Change to Transmission Charging Methodology—August 2013 (online). https://www.ofgem.gov.uk/press-releases/ofgem-recommends-change-transmission-charging-methodology. Accessed 03 Aug 2014.

Office for Gas and Electricity Markets [OFGEM]. 2014. (online). https://www.ofgem.gov.uk/. Accessed 18 Dec 2014.

Offshore Wind Cost Reduction Task Force. 2012. Offshore Wind Cost Reduction Task Force Report—June 2012.

Paterson, J. 2013. Energy Policy and Scotland's Constitutional Future—Setting the Scene: Presentation to Energy Policy and Constitutional Seminar, University of Strathclyde—Friday 18 January 2013 (online). http://www.scottishconstitutionalfutures.org/Portals/29/Paterson.pptx.

Pinsent Masons. 2012. New Renewable Support Bandings Give Certainty to Wind, Scottish Government Claims—14 September 2012 (online). http://www.out-law.com/articles/2012/september/new-renewable-support-bandings-give-certainty-to-wind-scottish-government-claims/. Accessed 20 July 2014.

Plant, G. 2013. Offshore Renewable Energy Developments in the British Islands: Legal and Political Risk. *Renewable Energy Law and Policy Review* 4: 189–228.

RenewableUK. 2014a. Press Release: RenewableUK Condemns Pickles' 50th Intervention in a Wind Farm Intervention (online). http://www.renewableuk.com/en/news/press-releases.cfm/renewableuk-condemns-pickles-50th-intervention-in-a-wind-farm-application. Accessed 03 Jan 2014.

RenewableUK. 2014b. Offshore Wind (online). http://www.renewableuk.com/en/renewable-energy/wind-energy/offshore-wind/. Accessed 15 Jan 2015.

Ross, A. 2012. Sustainable Development in the UK: From Rhetoric to Reality? Milton Park: Earthscan.

Scottish and Southern Energy. 2014. SSE Freezes Price Until 2016 (online). http://sse.com. Accessed 10 Dec 2014.

Scottish Government. 2007. The Government Economic Strategy (online). http://www.scotland.gov.uk/Resource/Doc/202993/0054092.pdf. Accessed 07 July 2014.

Scottish Government. 2010. Scottish Planning Policy—February 2010 (online). http://www.scotland.gov.uk/Resource/Doc/300760/0093908.pdf. Accessed 27 July 2014.

Scottish Government. 2011. 2020 Routemap for Renewable Energy in Scotland (online). http://www.scotland.gov.uk/Resource/Doc/917/0120033.pdf.

Scottish Government. 2012a. Marine Scotland Licensing and Consents Manual: Covering Marine Renewables and Offshore Wind Energy Development (Report R.1957) (online). http://www.scotland.gov.uk/Resource/0040/00405806.pdf. Accessed 20 Dec 2014.

Scottish Government. 2012b. Renewables Obligation Banding Review 2011–12: Scottish Government Response to the Consultation—13th September 2012 (online). http://www.scotland.gov.uk/Resource/0040/00401801.pdf. Accessed 15 Jan 2015.

Scottish Government. 2013. Letter to the Rt Hon Edward Davey MP, Secretary of State, Department of Energy and Climate Change (online). http://www.scotland.gov.uk/Resource/0043/00437662.pdf. Accessed 02 Jan 2015.

Scottish Government. 2014. Wind Turbine Appeal Decision Statistics—30 Tuesday December 2014 (online). http://www.scotland.gov.uk/Topics/Built-Environment/planning/Appeals/ourperformance/WindTurbineStats. Accessed 02 Jan 2015.

Scottish Government. 2015a. Marine Scotland (online). http://www.scotland.gov.uk/About/People/Directorates/marinescotland. Accessed 02 Jan 2015.

Scottish Government. 2015b. Renewable Energy: Marine (online). http://www.scotland.gov.uk/Topics/Business-Industry/Energy/Energy-sources/19185/17853-1. Accessed 12 Jan 2015.

Scottish Government. 2016. Energy Statistics for Scotland—June 2016 (online). http://www.gov.scot/Resource/0050/00502426.pdf. Accessed July 2016.

The Guardian. 2013. Renewable Generators Ask OFGEM to Cut Electricity Transportation Charges—11 July 2013 (online). http://www.theguardian.com/environment/2013/jul/11/renewable-ofgem-electricity-transportation-charges. Accessed 08 Feb 2015.

The Smith Commission. 2014. Report of the Smith Commission for Further Devolution of Powers to the Scottish Parliament—27 November 2014 (online). http://www.smith-commission.scot/wp-content/uploads/2014/11/The_Smith_Commission_Report-1.pdf.

Warren, C., and M. McFadyen. 2010. Does Community Ownership Affect Public Attitudes to Wind Energy? A Case Study from South-West Scotland. *Land Use Policy* 27: 204–213.

Wood, G. 2010. *Renewable Energy Policy in Scotland: An Analysis of the Impact of Internal and External Failures on Renewable Energy Deployment Targets to 2020—CEPMLP Energy Series*. Dundee: University of Dundee.

Wood, G. 2013. Connecting the Dots: A Systemic Approach to Evaluating Potential Constraints to Renewable Electricity Technology Deployment to 2020 and Beyond in the United Kingdom. PhD thesis, University of Dundee, 269.

Wood, G., and S. Dow. 2011. What Lessons Have Been Learned in Reforming the Renewables Obligation? An Analysis of Internal and External Failures in UK Renewable Energy Policy. *Energy Policy* 39: 2228–2244.

Author Biography

Dr. Geoffrey Wood is a Teaching Fellow in International Energy Law and Policy at Stirling University. An energy and environmental law, policy and regulatory specialist, his work focuses on the development of legal and governance frameworks for sustainable energy transitions to optimise policy delivery in terms of environmental, social, economic, political and technological outcomes. He has published over 30 articles, books, chapters and other publications. Currently also working as a Researcher and Contributing Lecturer at Dundee University, Geoffrey has previously held research, lecturing and consultancy positions for various organisations, including the Scottish Government and the Offshore Renewables Institute.

CHAPTER 3

Community Renewables: Balancing Optimism with Reality

Bill Slee and Jelte Harnmeijer

3.1 Introduction

Frequently lauded in the wider UK context, the evolving phenomenon of Scottish community energy is also increasingly receiving international attention, with the Scottish Government's flagship community energy fund recently highlighted by the Organisation for Economic Cooperation and Development (OECD) as a 'pioneering' example of bottom-up policy approaches to renewable energy (OECD 2012). Starting from a few scattered projects in the 1990s and early 2000s, the rate of uptake of Scottish community energy—both in raw capacity (kW) terms and in terms of absolute number of projects—has been roughly exponential, equivalent to a growth rate of almost 30% per year. However, this needs to be balanced against the observation that community-owned energy still contributes less than 4% of Scottish onshore renewable energy generation. In this chapter, we explore the context in which community ownership has developed, the rhetoric of community in Scottish policy, the wider regulatory and policy context, the roots of community energy policy, the diversity of practice, the implications of

B. Slee (✉) · J. Harnmeijer
The James Hutton Institute, Aberdeen, UK
e-mail: bill.slee@hutton.ac.uk

J. Harnmeijer
e-mail: Jelte.Harnmeijer@hutton.ac.uk

G. Wood and K. Baker (eds.), *A Critical Review of Scottish Renewable and Low Carbon Energy Policy*, Energy, Climate and the Environment,
DOI 10.1007/978-3-319-56898-0_3

increased devolution on community energy policy, the actual and potential benefits arising from community energy and the obstacles to increase its share of the renewable energy market in Scotland.

3.2 Context: History and Discourse

The evolving story of Scottish community energy cannot be seen in isolation from a widely articulated discourse about community empowerment that has evolved over the last 30 years and which has, although it extends beyond Scotland, taken a distinctly Scottish dimension. Although the dominant perspective of community empowerment in the latter half of twentieth-century Scotland related primarily to rural land in a challenge to Scotland's distinctly concentrated pattern of land ownership (Wightman 1996), by the second decade of the new millennium, the reach of Scottish-branded community empowerment had been extended to embrace much wider arenas of policy. Community empowerment, which first found legislative expression in the Land Reform (Scotland) Act 2003, is reiterated as a framing concept in the land use strategy and now underpins the Community Empowerment (Scotland) Bill which, after a substantial period of consultation, was introduced to the Scottish Parliament in June 2014 and passed into law in 2015. The Act deepens the opportunities for community ownership, creates opportunities for community ownership in towns, recognises the benefits of moving some land from public sector to community ownership and expands the scope for community acquisition of neglected or abandoned land. Further legislation is pending.

The rhetoric of community permeates Scottish political discourse and can be seen as part of a wider UK 'Big Society' discourse, but it has a quintessentially Scottish character. In a sense, the emergence of community energy as a facet of Scottish energy production is but one part of the warp and weft of community empowerment rhetoric and practice in the social and political life of Scotland. According to Becker and Kunze (2014), the idea of community energy is a distinctly UK conception, which in the rest of Europe may be better encapsulated in the broader terms '*collectively and politically motivated renewable energy*', a categorisation which includes the raft of local authority/municipal projects in many countries, all of which are essentially oppositional towards commercial corporate ownership of energy.

3.3 The Rhetoric of Community in Scottish Policy

At various times in the years after the Second World War, the use and purported misuse of rural land have been subjected to much critical scrutiny, although it is possible to trace the distant roots of community-based land reform back as far as the 1880s (Becker and Kunze 2014). In the last 30 years, the debate has intensified and resulted in significant government action. Issues relating to the use, misuse and underuse of land were debated in settings as diverse as government reviews, academic debate and radical theatre, the last most prominently in John McGrath's 7:84 Theatre Company. However, it was not until the late 1990s and New Labour's commitment to a Scottish Parliament, that community-based land reform was first actively discussed in government (Bryden and Geisler 2007), resulting in the formation of a Land Reform Policy Group in 1997, the report of which formed the basis of policies that were set in place in the 2003 Land Reform (Scotland) Act.

The core arguments for community-based land reform lay in the belief in creating a stronger voice for local community decision-making about land and the suggestion that some large landowners had failed to realise their land's productive potential. Modest communal decision-making about some aspects of land management had been enshrined in the crofting system, which dates in its present form from 1880s legislation. This, coupled with the strong sense of injustice that remained after the so-called Clearances of the nineteenth century, along with the evidence of land management practices that ranged from active stewardship through benign neglect to active sterilisation of development opportunities, created the preconditions for policy change in Scotland at a time when much of Europe was going through active decollectivisation policies (Swain 2007). The formalisation of land reform in legislation was preceded by a number of often publicly assisted community purchases of estates which had a history of poor landlord–tenant relations in places such as Assynt, Eigg and Gigha and more amicable handovers in places such as Borve on Skye. From the late 1990s, these activities were supported by advice and finance from a newly formed Community Land Unit of Highlands and Islands Enterprise.

On the back of the findings of the Land Reform Policy Group, the first Scottish Parliament passed the Land Reform (Scotland) Act in 2003, which created the opportunity for a government-assisted community right to buy, with stronger powers created in the crofting counties of the north and west which included the right to buy when a landowner had no wish to sell. Subsequently, the Forestry Commission responded to the wider debate about land ownership and a critique from bodies such as Reforesting Scotland of the Forestry Commission's management and disposal practices, by creating opportunities for communities to acquire ownership or management of local forests. Indeed, some of the earliest community land purchases were of Forestry Commission disposal sites, such as Abriachan in Inverness-shire in 1998. These could conceivably be supporting sustainable community renewable energy in the form of wood heat systems. To date, there is little evidence of community woodland owners looking to exploit the Renewable Heat Incentive (RHI) for community heating or indeed for supplying wood fuel to private owners who have used RHI support.

Rather than abating in the light of the combination of pioneering actions and ground-breaking legislation, the land reform movement has gained momentum, even if most of the exemplar sites and iconic examples date from the first flush of community acquisitions in the early 2000s. Since 2012, a new Land Reform Review Group (LRRG) has been working and it came up with substantive recommendations in 2014 (LRRG 2014). A consultation document in January 2015 sought views on extending the right to buy into urban communities and onto abandoned or mismanaged land (Scottish Government 2014). The Westminster Parliament's Scottish Affairs Committee also undertook a report on land reform reporting in March 2014 (Scottish Affairs Committee 2014). New land reform legislation was tabled in July 2015 following on from the work of the Land Reform Review Group. Community engagement in land and renewable energy is also evident in the land use strategy, which has its origins in a clause inserted into the Climate Change (Scotland) Act 2009. Page 27 of the strategy notes that: '*we are committed to maximising the opportunities for local ownership of energy as well as securing wider community benefits from renewables*' (Scottish Government 2011a). However, although the significant effort has been expended to increase the community benefit funds derived from commercial renewables, and in spite of some new policy means to support community energy, it would be hard to argue that other policies put in place have to date '*maximised*' opportunities for local ownership of renewable energy developments.

A further arena in which community empowerment has been exercised has been in community planning. Community planning was conceived more as a way of involving key agencies in ensuring joined-up delivery of public services and enabling a cascade of policy delivery from Scottish Government to councils to local communities. Community planning is intimately connected to the idea of partnership delivery of public services. The ministerial introduction of the discussion document relating to the bill asserts that community empowerment is at the core of a project 'about communities taking their own decisions about their futures' (Scottish Government 2013a). In 2013, the first minister promised a working group to explore greater devolution in the Scottish islands in the Lerwick Declaration, a promise that has now been broadened to include all Scotland's communities according to the press release on the Community Empowerment Bill (Scottish Government 2013b).

Notwithstanding these major developments, and given the Scottish Government's firm purpose statement, it is hard to see how a top-down 'managerialist' approach to policy delivery can be reconciled with the principles of more local '*self-determination, subsidiarity and local decision-making*' (Scottish Government 2013a: 2). As is evident in service delivery in general and in community energy policy in particular (see below), proposals for new styles of governance create scope for tension and disagreement both within communities and between local communities, across scales and between councils and central government. Community energy as an idea has caught on, but its implementation remains locked into energy governance and regulation designed principally for large-scale commercial corporate developments.

The link between community-based land reform, community planning and renewable energy production can be seen as part opportunistic and part a firm and literal assertion of the principles of community empowerment. A number of communities that had acquired land either in advance of or after the Land Reform (Scotland) Act in 2003 realised that delivering positive socio-economic outcomes from traditional land use was deeply challenging in many situations on poor-quality land in remote locations. It was easy to create a warm glow from community ownership but much harder to realise significant socio-economic improvements. UK renewables policy offered an opportunistic lifeline. The rapid increase in support for renewables through government-run financial subsidy mechanisms including the Renewables Obligation (RO) and, subsequently, Feed-in Tariffs (FiTs) created potential for high rates of return on renewables investments. Many of the early land reform activists were located in island and remote communities where there was often high technical potential for renewable energy production (even if grid connection was lacking). Revenue generated from renewables had the capacity to provide substantial injections into remote communities that were thereby able to insulate themselves from diminishing central government revenue streams. Local development could thus be shaped, to a greater degree, by local action and local income.

The scope for renewables developments on community-owned land can potentially be thwarted by an interposed lease. This is a legal arrangement put in place by a landowner prior to any community buyout to sell development rights of say, wind energy developments, to a third party in order to stop a successful community buyout gaining access to the renewables resource. This has happened in the case of Pairc estate on Lewis and deemed legal (Scottish Land Court 2007). Submissions to the most recent review of land reform have reflected on this issue.

3.4 The Wider Policy Environment

Policy to support community renewables is in a large part nested within wider energy policy, but also connects to a number of other areas of regulation and policy. In spite of strongly supporting policy rhetoric regarding community renewables, a case can be made that the policy architecture is rather less enabling than it might at first sight appear to be and, in some cases, might even be antagonistic to the development of community renewables. This argument stems from the observation that *ceteris paribus*, an individual community organisation, may be less able to plan, develop and deliver a given renewables project than a corporate energy company would be.

At issue is whether provision is being made for the fact that community projects differ from commercial 'analogues' in several important (and related) respects, including:

1. community projects typically take longer to develop;
2. community projects are often taken forward by non-specialist volunteers, rather than salaried professionals; and
3. community projects are often in a weaker position to secure debt finance.

Recent interventions by the Scottish Government have started to address the substantial handicaps in the realms of debt finance provision and a lack of access to project management expertise. Current policy to support community ownership of renewable energy production does not, however, extend to how community ownership is treated by the planning system, distribution and other grid network operators, and market-based renewable energy incentives (e.g. RO, FiTs, RHI and Contracts for Difference).

3.5 Planning and Statutory Consultations

The major regulatory hurdles for all renewables development proposals include planning permission and compulsory statutory consultation, such as the right to use water in the case of hydroelectric schemes. Both fall within the power of Holyrood rather than Westminster, and in neither case does special provision exist for projects brought forward by the community (as opposed to commercial, industrial or domestic) sectors. More particularly, positive externalities or other substantial socio-economic benefits that are unique to the community sector do not in practice constitute a material consideration in planning decisions.

The upshot is that at present, Scottish planning policy does not formally differentiate between a community-owned renewables scheme and a conventional commercial development application in how schemes are considered in the regulatory system. There exists some evidence, however, that community schemes enjoy a higher success rate in the planning system in practice (Haggett et al. 2013). Nevertheless, there is no functional mechanism within the planning system guidelines to obligate local authorities to assess trade-offs between local socio-economic benefit and environmental costs. And, given the complexities of local and global public goods and bads associated with renewables, expecting local councillors and officers to make well-informed judgements is a big ask. However, the town and country planning system is expected to make judgements based on the social, economic and environmental impacts of a proposal. With regard to renewable energy developments, guidelines are indicated in Scottish Planning Policy and are supplemented by online guidance sheets in relation to specific technologies (Scottish Government 2010). In practice, the treatment of social and economic impacts is rather stylised (into visual appraisal, noise assessment etc.), and socio-economic impact assessment is regarded by many planning experts as the weakest part of the environmental assessment process that routinely accompanies larger renewable energy proposals (Glasson and

Heaney 1993; Chadwick 2002; Slee 2013). All these have two crucial consequences. The first is that one of the major potential advantages—being able to deliver on local socio-economic benefit that community groups have over commercial players—is effectively muted. The second is that no formal incentives exist within the planning system to encourage commercial developers to partner with community groups.

In our view, it would be straightforward for the Scottish Government to make it easier for community energy to face reduced regulatory hurdles by asking planners to take stronger account of the widely acknowledged additional local economic benefits arising from community ownership. This is not an issue of taking ownership into account in the planning decision as some critics of such possible enabling powers suggest but an acknowledgement of the distinctive and much-enhanced beneficial impacts that community-owned renewables deliver locally.

3.6 Grid

A second area where community energy projects may face significant policy barriers is in grid connection. The policy surrounding grid connectivity is reviewed elsewhere in this volume (see the chapters by Toke and Wood). The grid regulator, OFGEM, is not directly accountable to Scottish Parliament, whilst UK policy has handed responsibility for grid management to the two main electricity producers in Scotland. UK practices have been described as 'opaque, onerous and inflexible' by industry bodies such as Renewable UK and Scottish Renewables. The community sector faces particular barriers relating to relatively small projects often in remote locations, and a recent report notes that: *'more work is needed to improve the transparency and predictability of grid connection processes and charges, to improve consistency and the communication channels between the Distribution Network Operators (DNOs) and generators'* (Cornwall Energy 2013: 4). In both Germany and Denmark, community projects have enjoyed priority access to the grid.

3.7 Market-Based Renewables Incentives

The nature of market-based incentives, such as FiTs, the RO and the RHI commonly set the 'bottom-line' that separates profitable projects from unbankable ones. As such, so-called degressions—periodic reductions in tariffs—represent hard deadlines by which a project must be licensed, if it is to lock into the higher pre-degression rate. In the complex and expensive world of renewables development, this translates directly to amplified risk for community projects. These risks are dramatically increased by DECC's 'consultation' in the summer of 2015 which signals the end of pre-accreditation, which means community groups would commit to construction with no knowledge of the tariff they would receive.

But there is another, more positive side. As off-the-shelf policy instruments, these incentive schemes also lend themselves well to enabling the uptake of community renewables. Examples of ways that they could be or have been employed to boost community energy sectors include premium rates for wholly and largely community-owned projects (e.g. Nova Scotia and Ontario), or the use of an elevated FiT cap as recently considered (but rejected) for the UK, or an extended pre-accreditation period specifically for community FiT registrations (accepted) (Department for Energy and Climate Change (DECC), 2014a).

3.8 EU State Aid Regulations and Tax Benefits

Many of the arguments we make in this chapter about regulatory barriers can also be brought to bear against the current EU legislative framework which, like key components of UK and Scottish policy, lacks explicit recognition and support for community renewables. A particularly serious and topical issue is the restrictive rules on State Aid, which constrain UK and other member states' ability to offer grant or 'soft loan' support towards covering capital costs for community renewables projects (European Commission 2014)

3.9 Investment Incentives

Tax benefits, such as the 'Enterprise Investment Scheme' (EIS) and the 'Seed Enterprise Investment Scheme' (SEIS), have played an important role in energising community investment into renewables. These benefits which have been, and are, crucial in securing investment into a growing number of Scottish community energy projects are currently under significant threat from new registration rules proposed by the Financial Conduct Authority (Financial Conduct Authority 2014).

The barriers faced by community energy developments are acknowledged by the Scottish Government, and some specific policies have been put in place now under the umbrella of the Community and Renewable Energy (CARES) scheme, which is considered in the next section after a brief historical overview.

3.10 The Roots of Community Energy Policy in Scotland

In some ways, practice has preceded policy in the recent development of community energy in Scotland. However, it is important to recall a longer history of communitarian energy policy in Scotland and the legacy of rural community empowerment that began much earlier. Tom Johnston, a radical Clydeside MP, who rose to become Secretary of State in the Second World War Coalition Government, pushed the development of hydroelectric power firstly in a 1943 Act establishing the North of Scotland Hydro Board and, after production passed into public ownership, expanded hydroelectricity production in one of the most sparsely populated and then impoverished rural communities in Britain. Hydroelectric developments not only provided a key service to rural areas but also, at the same time, the development and maintenance of the infrastructure created secure jobs. The privatisation of electricity production in the neoliberal sell-off of state-owned industry in the early 1990s supplanted the socio-economic logic that drove Tom Johnston with a much narrower market-driven logic. The re-engagement of communities at a more local level in the last decade represents a reincarnation of those communitarian principles which drove Tom Johnston in the post-war years, as acknowledged by the First Minister, Alex Salmond, when he spoke to the Community Land Scotland Conference in 2013 (Community Land Scotland 2013).

Recognition of the potential for community-owned renewable energy systems to support rural communities led to the formation of a unit within the Highlands and Islands Enterprise to support community renewables in 2004. The Highlands and Islands Community Energy Company (HICEC) provided advice and technical support to communities engaging in energy production from very small-scale installations in village halls to much larger schemes. In 2007, HICEC was offloaded from the public sector into a Scotland-wide company limited by guarantee called Community Energy Scotland, which remains active to the present.

As the public subsidy for renewables has grown in total volume (though not per MW of power produced) in pursuit of national and international targets for emissions reduction, so community groups have used renewables as a dual opportunity for generating community income, contributing to climate change mitigation and delivering wider community development aspirations. Indeed, a backward glance at many community land purchases reveals that, especially on the Inner and Outer Hebrides, the pursuit of renewable energy production has provided perhaps the most reliable and potentially lucrative financial return of any investment in their newly acquired landholdings. Where community buyouts are associated with community energy, the income stream was almost always diverted into additional local development activity, normally as a requirement of the trust established to manage the project. By such means, it was possible to break the negative cycle of what the New Economics Foundations describe as 'leaky bucket' economies in which the benefits of many natural resource-based economic activities pass quickly out of local areas to external shareholders (Entwistle et al. 2014; Phimister and Roberts 2012). Community-owned developments comprise a manifestation of the principle of community power for wider community empowerment.

The Scottish Government has consistently given support to community renewables. In its most recent statement in September 2015 (Scottish Government 2015a: 3), it notes Scotland's '*ambition to see community energy mainstreamed within a whole systems approach, with opportunity for community ownership and control across the full range of components in the system: generating low carbon energy, improving energy efficiency, distributing energy and even storing energy*'. It now has in place two support schemes for community renewables: CARES and the Renewable Energy Investment Fund (REIF). The CARES scheme was launched in 2011 and is a financial support package primarily to help overcome the high transaction costs and risks associated with negotiating a renewable energy project, repayable with interest if and when a scheme is approved, but not repayable should a project not get through the planning stage. It was initially managed by Community Energy Scotland, but following an open tendering procedure, management of the contract passed to a consortium of NGOs managed by the Energy Saving Trust (EST), who operate the scheme and provide other support services for community renewables as Local Energy Scotland. In 2013–2014, a separate stream of CARES funding was devoted to the Local Energy Challenge Fund, to support innovation in the sector, not exclusively with community-based enterprises. This led to 17 schemes receiving support.[1]

The CARES monies are now distributed through a family of funds. The £103 million REIF scheme was launched in 2012, in part to help meet the gap in the availability of debt finance for smaller (< £3m capital cost) projects. It was also set up to help address finance gaps encountered by private or community developers looking to put forward newer renewables technologies, focussing on marine and community heating schemes. It is managed by the Scottish Investment Bank and has proved a flexible and instrumental tool in providing finance to a variety of community projects. Examples include a 2013 loan of £49,000 to Gigha Green Energy Ltd, a subsidiary of the Gigha Development Trust, to help meet a funding shortfall on an additional turbine added to three existing community turbines; a loan towards a 9 MW 100% community-owned wind farm developed by the Point and Sandwick Development Trust on Lewis; and a loan to support a 25% stake in a 3-turbine wind farm in South Lanarkshire. The majority of REIF funds, however, have been directed to private sector offshore renewables developments.

Either through intent or otherwise, the Scottish Government has rather fudged the furtherance of community renewables. This has happened in at least two ways. First, the term 'community benefit' has been used in Scotland to describe payments made by commercial operators to communities as well as the benefits to communities arising from *their own* investment in renewables or co-investment by communities and private operators. The gulf in community receipts between a community-owned renewable energy scheme, which might be expected to generate between £130,000 and £200,000 of income per MW/year net of borrowing, and a community benefit scheme paying typically £3000 per MW/year is immense. It is difficult not to get the impression that the fudging between full community participation in renewable energy and what one community activist has termed the colonialist 'beads and necklaces' approach of large-scale corporate investors paying off the local population is intentional and comprises a pragmatic response by government to ensure major developers feel unthreatened by the rhetoric of community used in debate about energy production and more widely.

A further fudging of the notion of community is evident in the conflation of community- and locally owned renewables within the Scottish Government's target of 500 MW of community- and locally owned capacity by 2020. Indeed, in early October 2015, the Scottish Government announced that it had achieved its community energy target 5 years early (Scottish Government 2015b), when in practice, only about 70 MW of community energy capacity exists in Scotland. The target of 500 MW is commonly referred to as a 'community renewables target', but includes the crucial term 'locally owned', thus diluting the target through incorporation of a much larger project portfolio developed by for-profit rural businesses. In consequence, the community component actually represents a minority of the current community- and locally owned category. Why this might matter is that with community ownership, there is a virtual certainty that the proceeds will be reinvested locally, but with private ownership, there is absolutely no such guarantee nor indeed a solid evidence base to suggest that high levels of local reinvestment do occur from private owners, although circumstantial evidence reveals some farmers recapitalising farms on the back of renewable energy income streams (Mackie 2015).

Community shareholding has been seen in the rest of the UK as a potential mechanism to engender more local support for renewables developments. Indeed, it has been argued that public sentiment recognises the need for social justice in renewables developments (Cowell et al. 2011). Recent policy developments in the 2015 Infrastructure Act in England and Wales will create a time-lagged backstop measure to obligate developers to offer community shareholdings in renewables projects of above 5 MW, unless there is compelling evidence that the utility companies are already delivering community co-ownership voluntarily. Furthermore, the UK strategy paper on community energy reinforces the case for community shareholding (DECC 2014b).

Energy co-operatives have also recently gained momentum in Scotland; examples of projects completed or under construction include Dingwall Wind Co-operative, Garmony Hydro, Harlaw Hydro, Islay Energy (Wind) and Wester Derry (Wind), with several others in the pipeline. More widely in its policy documentation on renewables, the Scottish Government reiterates the case for community ownership. Pledge No 1 of the Climate Change Delivery Plan (Scottish Government 2009a: 8) asserts: *'we will support and accelerate the implementation of renewable energy, through our Renewable Energy Action Plan, in a way which promotes large-scale, community based, decentralised and sustainable generation'*. Part of the vision of the Renewables Action Plan (Scottish Government 2009b: 48) is *'to maximise the benefits for rural communities from renewable energy, not only in terms of access to locally produced low carbon energy but in terms of social cohesion and economic development'*. The Scottish Government's Routemap for Renewable Energy in

Scotland (Scottish Government 2011b) asserts that '*we wish to maximise the benefits for communities from renewables and to transform the level of opportunity for local ownership of energy. Our ambition is for all Scottish communities to share in the rich rewards of our next energy revolution*'. This policy is reiterated in the Electricity Generation Policy Statement of 2012 and in the Scottish Government's Report on Policies and Proposals 2 (Scottish Government 2013c: para 4.2.1).

In summary, although the rhetoric of community empowerment is especially strong in Scotland, in relation to renewable energy, there is arguably less support for differentiating community renewables from renewable energy developments more widely than there is in the rest of the UK. Having set a target of 500 MW of community renewables output by 2020, the Scottish Government later adjusted this target to 500 MW of community- and locally owned renewables. It has also conflated the benefits which a community receives from private companies under what are essentially planning gain arrangements (which typically range from £2000 to £5000 per MW per year). The greater adoption of community renewables may be more a function of exploitation of opportunities by communities energised by community-based land reform. Additionally, such an approach might help overcome the antagonism of many rural residents to wind turbines, the predominant, but by no means only technology used to date.

However, post-2015 UK election moves to reduce support for renewables at the UK level represent a sea change in policy which will have devastating consequences on renewables generally and community energy in particular. Realistically, the recent Westminster policy changes signal the death knell of new community renewables projects in Scotland, unless additional Scotland-specific support measures are put in place. What was written as a manifesto by the Scottish Government as recently as September 2015 is, in reality, more an epitaph, with the Scottish Government currently seeking ways to salvage what it can from the wreckage of the UK renewables support architecture.

3.11 The Diversity of Community Energy Practices in Scotland

There are currently reputedly 360 community renewable energy projects in Scotland with a capacity of just over 30 MW of installed operational capacity (Haggett et al. 2013). An attempt was made to map these for DECC in 2012 and more recently by an international consortium

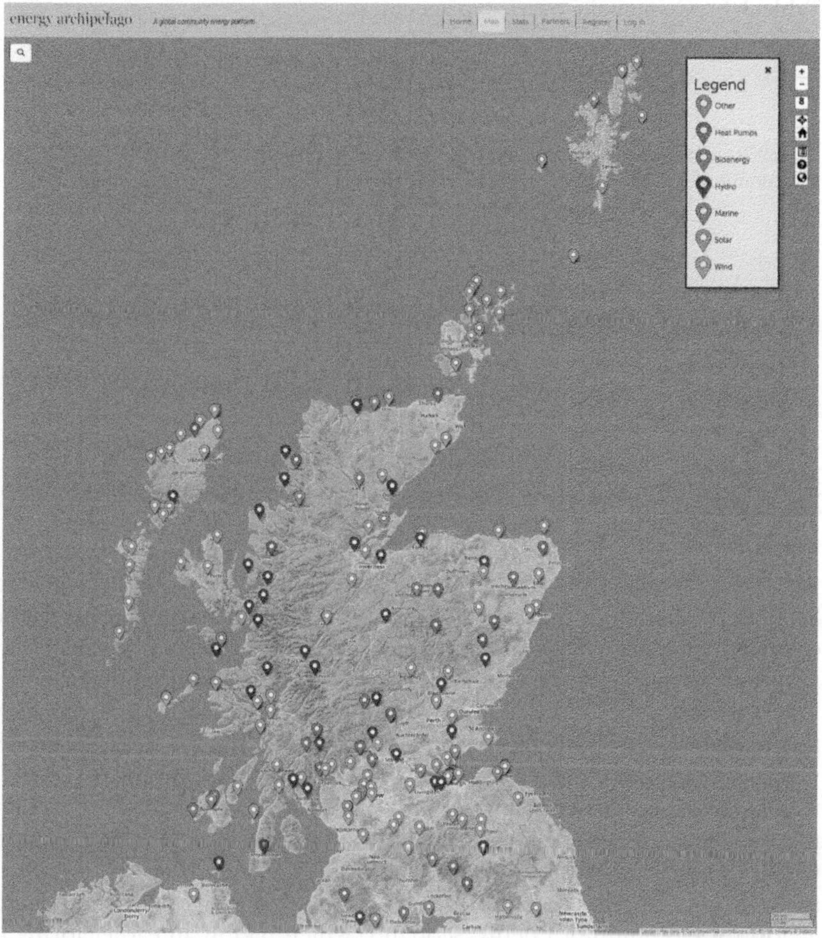

Fig. 3.1 Map of Scottish community energy projects. *Source* Energy Archipelago February 2015

of research and practitioner organisations under the project name of 'Energy Archipelago' (see Fig. 3.1).

We are not aware of any other country in the world where such a wide diversity of business models and legal structures are used for community energy.[2] In other words, there is a myriad of different ways in which

Table 3.1 Community ownership in context

		A. Project ownership	
		Whole	*Partial*
B. Investment source for community stake	Community body *includes local development organisations such as development trusts*	Community-led projects	Revenue-sharing arrangements with local developments organisations joint ventures
	Individuals *includes co-operatives of all kinds*	Wholly cooperatively-owned projects Wholly crowd-funded projects	Revenue-sharing arrangements with co-operatives

Scottish communities can have a stake in renewable energy production. Table 3.1 shows an attempt to summarise them.

One model—that of a community-based local development organisation—usually taking the legal form of a development trust—has dominated Scottish developments to date. In both the European and UK context, the dominance of this particular model is a uniquely Scottish phenomenon.[3] Its preponderance may be more a consequence of the institutional support for such approaches rather than any intrinsic merit (Toke and Harnmeijer, forthcoming). A second approach is to use a co-operative business model which is much more common in many other European countries.

The development trust model has been widely used in local development projects in Scotland, especially in the Highlands and Islands region which can be seen as the cradle of community energy projects, largely on account of a superior renewables resource base, as well as the supporting actions of Community Energy Scotland, which had in an earlier form, as HICEC, supported community energy only in the Highlands and Islands Enterprise (HIE) area. Arguably, the explicit social remit of HIE created a sharper focus on community business models suited to the needs of the region. This model is highly suited to place-based local development where a group of informed local citizens provide the leadership and manage the disbursement of funds in accordance with the specification of the local trust articles.

The second model that of an energy producers' co-operative has also been widely used as a business model in community energy globally, in

Europe and in the rest of the UK, but has been sparingly used to date in Scotland. In a co-operative model, the enterprise is owned collectively by the membership, but whereas in a private company the number of shares shapes the power of an investor, a co-operative is based on each member having equal voting rights regardless of their investment. The membership can come from a geographically defined area, from a group of interested membership not defined geographically or most typically from a hybrid between the two. The Baywind Co-operative in Cumbria pioneered co-operative ownership of renewable energy in the UK in 1996, building on a business model widely used in Scandinavia and now promoted actively by Energy4All, an offshoot of Baywind. There are four Energy4All supported projects in Scotland and other co-operative renewables enterprises, including Dingwall Wind Co-operative and a number of hydroelectric projects.

The commercial-community partnership model of ownership, which was being actively promoted by DECC and in the UK Infrastructure Act and by the Scottish Government in its recent Community Energy Policy Statement Draft, has been developed in a number of cases where mainstream commercial developers have entered into some form of revenue sharing arrangement with a community. Many models exist. The initial development risk may have been carried by a commercial developer, which subsequently transferred ownership of a shareholding in a development vehicle to a community entity, or a community may have promoted a local scheme and drawn in a commercial renewables company as a partner, as at Neilston in East Renfrewshire. The Fintry model in Stirlingshire is perhaps the best known Scottish example, where a handful of members of an engaged small community contemplating a stand-alone renewables development instead worked with a commercial developer to obtain a shareholding in a bigger commercial development in 2007. Neilston Development Trust (based near Glasgow) has also developed a joint venture model between a commercial developer and the local community. The commercial-community partnership model also operates with a number of community co-operatives in Golspie in the Highland region and at Boyndie in Aberdeenshire.

The increased use of crowd sourcing of funds has been used in many arenas, including in renewables, as by Abundance Generation and Ecotricity. Although it potentially democratises ownership of renewable energy production, it is a democratisation for those with free resources to invest and unlikely to engage the fuel or energy poor.

As well as considering ownership structures and business models, it is also important to recognise that community groups may need to find novel ways to circumvent some of the obstacles they face. This may create new and beneficial forms of linkage between community energy and local communities. Many community groups face major grid connection issues. One way around this is to consider energy storage or off-grid local sales of energy. Some CARES-funded initiatives are exploring such issues. Two community hydroelectric schemes, at Kingussie and Applecross, are looking to sell electricity to a local source of demand in an off-grid sale in one case and to connect into a community heating scheme in the other.

3.12 The Implications of Increased Devolution on Community Renewables in Scotland

We argue above that the biggest current constraint on the expansion of community renewables in Scotland, until the recent change in FiTs and pre-accreditation, was the planning system's unwillingness to consider properly the widely acknowledged local socio-economic benefits arising from community ownership (in contrast to 'traditional' commercial ownership) in planning decisions and guidance. To change this is entirely within the Scottish Government's control, as planning is a fully devolved matter. More recently, the new constraints of lower FiTs and loss of pre-accreditation will undoubtedly reduce the scope for community engagement. Furthermore, if renewables production is pushed offshore, it is more difficult to argue a social justice case for community engagement with underwater or offshore wind turbines as there is often no obviously adversely impacted community, although community benefit funds are mooted. We note, however, that the offshore Middelgrunden wind farm in Denmark, 3.5 km from Copenhagen, has a significant component of community ownership, so an obligation on offshore renewables developers to offer coastal (or wider) communities a shareholding might capture some of the benefits for such places.

It is challenging to speculate on the impact of increased devolution on community renewable energy production. Scotland has a synergistic relationship with the rest of the UK, in that it has high potential for renewables and can draw on a UK-wide consumer subsidy to develop the Scottish renewables industry. If the link with UK energy markets and the associated UK-wide consumer subsidy of renewables were lost, energy bills would rise markedly in Scotland, as approximately 90% of the renewables subsidy to Scottish renewables comes from outwith Scotland and this subsidy would most likely have to come from Scottish consumers if Scotland decoupled from UK support systems.

Scottish renewables developments are highly dependent on rates of financial support and long-term political support for different renewables technologies (and scales) again determined outwith Scotland, albeit in consultation with Scottish authorities. The reductions in FiTs for onshore wind, hydro and solar, announced on the 27 August 2015 are already having a strong deterrent effect on developments in Scotland. As stronger anti-renewables sentiments have prevailed at UK level, the impact on Scottish renewables developments will almost certainly be negative. On the other hand, had English resistance to onshore wind been countered by a policy of forcing commercial developers to offer shareholdings to local communities, imitative policies in Scotland which forced developers' hands could have had a beneficial effect on community shareholding in the renewables sector.

Scottish community renewables developments are also contingent on grid connection. In a post-referendum paper, Scottish Renewables argue the case for stronger support for grid connectivity to major islands where so much of the community renewable activity has been based (Scottish Renewables 2014). Greater devolution of the management of the grid connection and giving greater weight to grid connections for community projects would be enormously beneficial for the development of community renewables, so too would be additional support for community-based local grid projects or linked district heating schemes.

It seems likely that an in-depth stocktaking will be required, building on the experience of policy turmoil and trying to map a way forward. Genuine community energy proposals may find funding from other sources, such as European structural funds, and farm level schemes could be funded from the Scotland Rural Development Programme (RDF). Perhaps the best that can be hoped for is initiatives such as the Local Energy Challenge Fund that funds innovative initiatives that are finding ways to overcome the present problems, for example, through local grid, CHP schemes and hydrogen production.

3.13 The Benefits Arising from Community Energy in Scotland

Slee (2015) identified seven main benefits attributable to community ownership of renewables, namely reduced opposition, increased local economic impacts, environmental justice, wider energy emissions reductions, reduced fuel poverty, filling the gap caused by local authority funding cuts and overall enhanced resilience of communities.

Co-ownership could reduce the opposition of some people within the recipient community to renewables, especially onshore wind, even if it is unlikely to appeal to those critics fuelled by an almost visceral opposition to wind power. If this were the case, it would help inculcate a wider 'pro-renewables' culture, thereby reduce development risk for community and commercial projects alike and, at the same time, assist the Scottish Government meet its statutory targets. Where there is clear evidence of positive income streams entering rural communities of potential reinvestment of these in energy-saving measures, local amenity and community capacity building, it is likely that at least some people who are 'sitting on the fence' may be persuaded of the merits of onshore wind (Haggett 2004).

It has been argued that '*local investment also provides an opportunity to strengthen and diversify local economies, particularly in rural areas, and can lead to new projects through the sharing of information and relevant experiences*' (Sawin 2004). The disparity between the injection into local economies from community and corporately owned wind farms is enormous. These benefits can be explored in terms of direct financial returns or in terms of wider economic benefits including multiplier effects. Co-ownership or shared equity has been initiated in a number of places in Scotland. Economic modelling work by Strathclyde University's Fraser of Allander Institute on Shetland has shown how co-ownership of a major onshore wind farm would have significant beneficial effects on the local economy, whereas external ownership would leach most of the benefits out of the rural community (Allen et al. 2011). Their findings are supported by recent studies in rural Wales which report very modest beneficial impacts on rural economies of large-scale externally owned wind farms due to the size and nature of community benefits received and how these are being utilised (Munday et al. 2011). The impact on local incomes can be very significantly enhanced by co-ownership, but the aggregate effect of enhanced local incomes depends both on what any income is spent on and where it is spent (Phimister and Roberts 2012).

The people most adversely affected by the amenity intrusion of an onshore wind development are the adjacent local communities and may consider that they merit environmental justice. In environmental planning, there has long been the principle of an environmentally compensating project, such that, for example, if a bypass destroys some ancient woodland, the developer agrees to new native woodland planting somewhere else locally. It has been argued that '*providing benefits to communities affected by wind-farm development is a matter of justice: a means of redressing the impacts on communities adversely affected by wind farms*' (Cowell et al. 2011: 1). Where the adversely affected party is the local community, the enhancement of village amenities and green space through enhanced local spending comprises a de facto community-compensating project for the visual intrusion of wind energy installations. One energy company operating in Scotland offers reduced bills to households adjacent to developments. However, there is a tension between communities feeling that they are 'bribed' to accept wind energy developments and a strong feeling that local people should benefit from such developments (Cass et al. 2010). It is, however, apparent that the income streams arising through ownership or co-ownership are up to tenfold over what commercial developers are currently offering in contributions to community funds.

A community that is empowered (literally) by co-ownership of an onshore wind (or other renewable energy technology) may engage more fully with the need to reduce other emissions and to keep Scotland as a leader in the struggle against the adverse global impacts of human-induced climate change. Further research is needed on whether communities that are engaged on energy issues and have built capacity to collaborate are likely to be at the forefront of post-carbon communities. The Transition Towns movement (Hopkins 2008) provides an illustration of multifaceted responses to climate change, though robust analysis may still be lacking.

Full or partial community ownership provides communities with funds which can be redeployed to address energy or other social and economic issues in local communities. To date, there have been no attempts to formally hypothecate revenue streams to deal with, for example, fuel poverty. However, there are examples of strategic expenditure by some community energy projects to address fuel poverty, for example, in the investments in insulation and home improvements on Gigha, an island off the west coast of Scotland where the community has invested significantly in onshore wind energy.

As a result of the very tight financial situation globally, the Scottish Government has found it necessary to reduce local authority budgets. The result is that services have been cut. If significant income is made available to communities through outright ownership or equity in renewables installations, such funds can help to plug the gap arising from reduced public expenditure. It is thus clearly in the government's interest to nurture the development of alternative funding streams for local projects. This availability of funding provides a platform for the actions of empowered communities, which also resonates strongly with Scottish Government policies.

Finally, Harnmeijer et al. (2012) also argue that local ownership delivers greater resilience and that communities can be a source of investment capital, though the latter is more likely under co-operative models rather than the community development trust model. Many rural economies are relatively undiversified economies, and collaborative effort in energy production can strengthen community resilience and economic resilience simultaneously.

3.14 The Obstacles to the Further Development of Community Renewables in Scotland

There are many obstacles to the further development of community renewables (see Fig. 3.2). These include the lack of skills and local capacity for specialist tasks such as project management, the weak degree of subsidiarity in Scottish and UK local government, the strong negativity arising from the recent changes in the Westminster-driven renewables policy agenda and the failure to ensure ready access to grid and the failure of the regulatory system to give due recognition to the added value of community ownership. Scottish Government policy is addressing the skills gap through CARES and Local Energy Scotland who are assisting *inter alia* by setting up framework contracts through which project management, legal and financial services can readily be procured.

The degree of local decision-making in both Scotland and the UK more widely is markedly less than in a French commune (village) or a United States (US) or Canadian small town. For all the rhetoric of local empowerment, there is not much of it in either Scotland or the UK

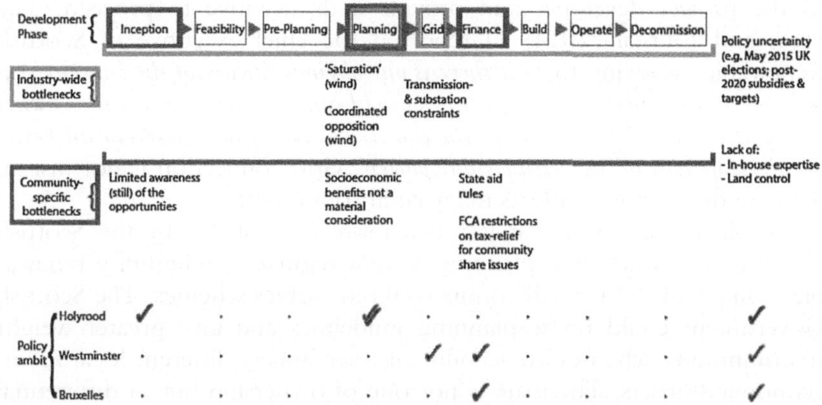

Fig. 3.2 Schematic overview of obstacles facing Scottish community renewables, mapped on to relevant 'policy ambit'

more widely. It is most strongly expressed where active communities with strong social capital and wider skills have been instrumental with agency and Scottish Government support in community land buyouts. Many successful renewables projects have been developed by such communities. It takes skills, grit and major effort to deliver a significant community renewables project. Furthermore, if grid connection (and wider relocalisation thinking) makes it incumbent on communities to devise more elaborate schemes (such as linked district heating schemes or local grids), the technical, financial and project management needs will escalate accordingly.

It is likely that the shift of support away from onshore wind at the UK level will deter community schemes, for onshore wind remains the most cost effective technology for most communities. Uncertainty can have a corrosive effect on confidence and the current UK Government's deep antagonism to onshore wind does little to encourage project development. A community-specific FiT would be a potential way forward, but the UK government showed more inclination for a co-ownership model and then proposed neither. However, as the UK government opts for supporting higher cost renewables, this itself will raise prices and may further diminish public enthusiasm for the renewables project.

The grid connection issue is critical. When FiTs are falling over time, delay in getting a project up and running can entail a substantial cost to the project developer and provide a disincentive to proceed. The Press and Journal (2014: 1) reported the chief executive of Scottish Renewables as saying that *'if there is one obvious failure of the current regulation of our industry it is the lack of grid connections to Shetland, Orkney and the Western Isles—home to the country's best wind resources and key to the development of wave and tidal power'*. Grid connection is not just an issue on the islands; it affects many rural communities.

The final constraint that is much more controllable by the Scottish Government is how the planning system regulates community renewables compared to 'normal' commercial renewables schemes. The Scottish Government could revise planning guidelines and give greater weight to community schemes on account of their hugely different local socio-economic impacts. The issue is not one of ownership but of differential beneficial impact.

However, it may be possible to remove one constraint—the planning system—which is in the hands of the Scottish Government and then be faced with other constraints that are at present almost wholly outwith the Scottish Government's power to control, because they are Westminster decisions, albeit made in consultation with the devolved governments. Nevertheless, although the SNP achieved a landslide victory in the May 2015 general election, winning 56 seats (out of 59), it faces opposition from a majority Conservative government on almost all of its policy objectives, and so whether it will exercise its powers to support community renewables developments remains a moot point.

3.15 Conclusion

To conclude, we would recommend that Scottish policy for renewables be recalibrated to provide a much more explicit legal basis for consideration of the additional benefits of community energy over other ownership types. Such reform would represent recognition of the unique and currently undervalued benefits that community energy delivers, and thereby ensure that the design of Scottish policy more fully realises the pathway to energy decarbonisation and emissions reduction objectives. Some of the more important purported benefits associated with community ownership take time to materialise (see Fig. 3.3). It is worth positing that there thus exists less of an incentive for policymakers to capitalise on benefits that will only materialise in future election cycles—a 'democratic premium' that famously afflicts many facets of climate change policy.[4]

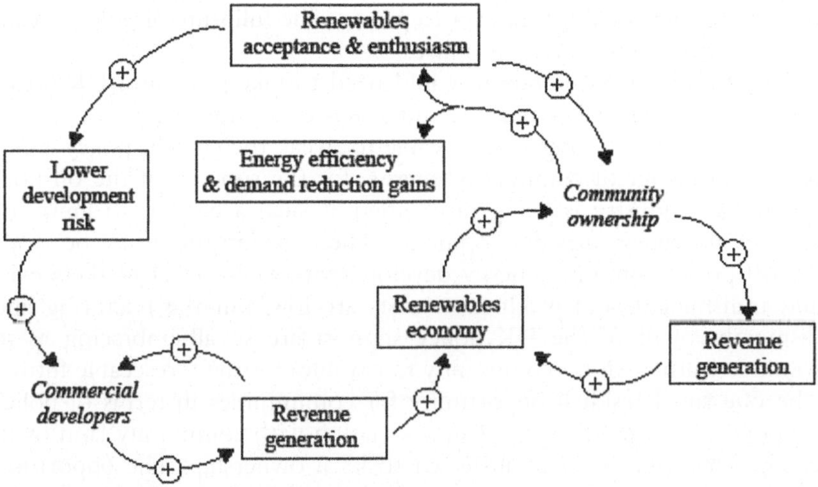

Fig. 3.3 Selected benefits associated with community renewables ownership, and feedback effects on the wider economy

Individual community renewables projects tend to take longer to complete and are typically of smaller scale, than commercial analogues (Harnmeijer et al. 2015). This may provide a disincentive to promote the sector: for instance, governments with ambitious and highly politicised decarbonisation, and emissions reductions targets may be tempted to opt for a 'big bang' large-scale commercial roll-out approach, at the cost of bottom-up grassroots-focused policy. Arguably, this is exactly what happened with the fraught tendering process for renewables development rights on Scottish Forestry Commission land, which saw development rights apportioned to a small group of very large companies.

Some policy developments which have taken place have provided greater incentives and support, and these are to be applauded. However, the post-referendum settlement which retains the Scottish Government's capacity to support community renewables may fail to realise its promise because of the failure to explicitly recognise in the regulatory system the considerable benefits that community ownership or co-ownership creates. The devolved nature of planning does create scope for the Scottish Government to prioritise community renewables and back the rhetoric with policy means. But it has yet to provide the full suite of policy means to deliver the extraordinary promise.

The called for recalibration would need to take place in a UK policy environment which has become much more negative since May 2015. Arguably, it will no longer be possible to tweak the current policy architecture in favour of community energy, for the support of the onshore renewables industry is being dismantled at such a pace as to limit any capacity for renewables developments. There may in the future be scope for off-grid community energy developments where local markets exist and transmission and production costs are low, but the fracturing and dismemberment of the UK policy support are so all-embracing as to sound a death knell for community renewables for the foreseeable future. The one small residual opportunity for communities in terms of policy support is the use of the RHI in association with community land ownership where forests are transferred to local ownership. The opportunities here remain considerable, but the actual developments to date are very limited. In the event of a turnaround of policy, communities need to be prepared, but borrowing from Robert Burns, *'an' forward, tho' we canna see, (we) guess an' fear!'*

Notes

1. An equivalent English scheme the Local Energy Assessment Fund launched in 2011 did deal exclusively with community groups.
2. For an international review, see Haggett et al. (2014).
3. Legal structures similar to 'Development Trusts' are used in community energy project further afield, such as in South Africa.
4. For a good discussion, see Gardiner (2013), A Perfect Moral Storm: The ethical tragedy of climate change. Oxford University Press

References

Allen, G., P. McGregor, and J.K. Swales. 2011. The Importance of Revenue Sharing for the Local Economic Impacts of a Renewable Energy Project: A Social Accounting Matrix Approach. *Regional Studies* 45 (9): 1171–1186.

Becker, S., and C. Kunze. 2014. Transcending Community Energy: Collective and Politically Motivated Projects in Renewable Energy (CPE) Across. *People, Place and Policy* 8 (3): 180–191.

Bryden, J., and C. Geisler. 2007. Community-Based Land Reform: Lessons from Scotland. *Land Use Policy* 24: 24–34.

Cass, N., G. Walker, and P. Devine Wright. 2010. Good Neighbours, Public Relations and Bribes: The Politics and Perceptions of Community Benefit Provision in Renewable Energy Development in the UK. *Journal of Environmental Policy & Planning* 12 (3): 255–275.

Chadwick, A. 2002. Socio-Economic Impacts: Are They Still the Poor Relations in UK Environmental Statements? *Journal of Environmental Planning and Management* 45 (2): 3–24.

Community Land Scotland. 2013. *Community Land Ownership—A Fairer Scotland: Conference Report: Appendix C—Alex Salmond, Keynote address at the Community Land Scotland Annual Conference*. Skye (online). http://www.communitylandscotland.org.uk/wp-content/uploads/2014/04/Community_Land_Scotland_Conference_Report_-_Annual_Conference_2013.pdf.

Cornwall Energy. 2013. *Overcoming Grid Connection Issues for Community Energy Projects a Report for Co-operatives UK and the Co-operative Group* (online). http://www.localenergyscotland.org/media/33075/Cornwall-Energy-report_Overcoming-grid-connection-issues-for-community-energy.pdf.

Cowell, R., G. Bristow, and M. Munday. 2011. Acceptance, Acceptability and Environmental Justice: the Role of Community Benefits in Wind Energy Development. *Journal of Environmental Planning and Management* 54 (4): 539–557.

DECC. 2014a. *Government Response to the Consultation on Support for Community Energy Projects Under the Feed-in-Tariffs Scheme—URN:14D/387 November 2014* (online). https://www.gov.uk/government/uploads/system/uploads/attachment_data/file/374540/Govt_Response_to_community_FITs_consultation_-_FINAL.pdf.

DECC. 2014b. *Community Energy Strategy: Full report—27 January 2014* (online). https://www.gov.uk/government/uploads/system/uploads/attachment_data/file/275163/20140126Community_Energy_Strategy.pdf.

Energy Archipelago. 2015. *The Global Community Renewables Portal* (online). https://energyarchipelago.com/.

Entwistle, G., D. Roberts, and Y. Xu. 2014. *Measuring the Local Economic Impact of Community-Owned Energy Projects (Scotland) Commissioned by—Community Energy Scotland, Gilmorton Rural Development and the James Hutton Institute (Aberdeen)* (online). http://www.communityenergyscotland.org.uk/userfiles/file/stevens_uploaded_documents/Measuring-the-Local-Economic-Impact-of-Community-Owned-Energy-Projects-2014-pdf-logo.pdf.

European Commission. 2014. *Communication from the Commission: Guidelines on State Aid for Environmental Protection and Energy 2014–2020, SWD (2014) 140* (online). http://ec.europa.eu/smart-regulation/impact/ia_carried_out/docs/ia_2014/swd_2014_0139_en.pdf.

Financial Conduct Authority. 2014. *Guidance on the FCA's Registration Function Under the Co-operative and Community Benefit Societies Act 2014—October 2014: Consultation Paper CP14/22* (online). https://www.fca.org.uk/static/documents/consultation-papers/cp1422.pdf.

Gardiner, S.M. 2013. *A Perfect Moral Storm: The Ethical Tragedy of Climate Change*. Oxford: Oxford University Press.

Glasson, J., and D. Heaney. 1993. Socio-Economic Impacts: The Poor Relations in British Environmental Impact Statements. *Journal of Environmental Planning and Management* 36 (3): 335–343.

Haggett, C. 2004. *Tilting at Windmills? The Attitude–Behaviour Gap in Renewable Energy Conflicts*. Final Report of ESRC Environment and Behaviour Programme Project (online). http://www.psi.org.uk/ehb/docs/finalre-port-Haggett.pdf.

Haggett, C., C. Creamer, J. Harnmeijer, M. Parsons, and E. Bomberg. 2013. *Community Energy in Scotland: The Social Factors for Success*. University of Edinburgh, ClimateXChange report (online). http://www.climatexchange.org.uk/files/4413/8315/2952/CXC_Report_-_Success_Factors_for_Community_Energy.pdf.

Haggett, C., M. Aitken, D. Rudolph, B. van Veelen, J. Harnmeijer, and M. Markantoni. 2014. *Supporting Community Investment in Commercial Energy Schemes* (online). http://www.climatexchange.org.uk/reducing-emissions/supporting-community-investment-commercial-energy-schemes/.

Harnmeijer, A., J. Harnmeijer, N. McEwen, and V. Bhopal. 2012. *A Report on Community Renewable Energy in Scotland*. Scene Connect Report Edinburgh.

Harnmeijer, J., A. Harnmeijer, V. Bhopal, S. Robinson, E. Phimister, D. Roberts, and J. Msika. 2015. *The Comparative Costs of Community and Commercial Renewable Energy Projects in Scotland*, ClimateXChange (online) http://www.climatexchange.org.uk/reducing-emissions/comparative-costs-community-and-commercial-renewable-energy-projects-scotland/.

Hopkins, R. 2008. *The Transition Handbook: From Oil Dependency to Local Resilience*. Totnes: Green Books.

House of Commons, Scottish Affairs Committee. 2014. *Land Reform in Scotland: Interim Report, Eighth Report of Session 2013–14*. Report, together with formal minutes Ordered by the House of Commons to be printed 18 March 2014.

Land Reform Review Group. 2014. *The Land of Scotland and the Common Good*. Final report of the Land Reform Review Group.

Mackie, B. 2015. *Presentation to North East Scotland Agricultural Advisory Group*, 28 Jan 2015 (online). http://committees.aberdeenshire.gov.uk/FunctionsPage.aspx?dsid=82020&action=GetFileFromDB.

Munday, M., G. Bristow, and R. Cowell. 2011. Wind Farms in Rural Areas: How Far Do Community Benefits from Wind Farms Represent a Local Economic Development Opportunity? *Journal of Rural Studies* 27: 1–12.

Organisation for Economic Cooperation and Development (OECD). 2012. *Linking Renewable Energy to Rural Development* (online). http://www.oecd-ilibrary.org/urban-rural-and-regional-development/linking-renewable-energy-to-rural-development_9789264180444-en.

Phimister, E.C., and D.J. Roberts. 2012. The Role of Ownership in Determining the Rural Economic Benefits of On-Shore Wind Farms. *Journal of Agricultural Economics* 63 (2): 331–360.

Sawin, J. 2004. *National Policy Instruments: Policy Lessons for the Advancement & Diffusion of Renewable Energy Technologies Around the World*. Thematic Background Paper, Secretariat of the International Conference for Renewable Energies, Bonn.

Scottish Government. 2009a. *Climate Change Delivery Plan: Meeting Scotland's Statutory Climate Change Targets—June 2009*.

Scottish Government. 2009b. *Renewables Action Plan—June 2009*.

Scottish Government. 2010. *Scottish Planning Policy*. Scottish Government.

Scottish Government. 2011a. *Getting the Best from Our Land: A Land Use Strategy for Scotland*.

Scottish Government. 2011b. *2020 Routemap for Renewable Energy in Scotland*.

Scottish Government. 2013a. *Consultation on the Community Empowerment (Scotland) Bill*.

Scottish Government. 2013b. *New Powers for Scotland's Communities: Plans for Greater Local Decision Making*.

Scottish Government. 2013c. *Low Carbon Scotland: Meeting the Emissions Reductions Targets 2013–2027—The Second Report on Proposals and Policies*.

Scottish Government. 2014. *A Consultation on the Future of Land Reform in Scotland*.

Scottish Government. 2015a. *Community Energy Policy Statement Final Version published September 2015*.

Scottish Government. 2015b. *Community Renewables Meets Target Early: Estimated 508 MW in Community and Local Ownership*.

Scottish Land Court. 2007. *Scottish Ministers vs Pairc Trust Limited and Others* (Application RN SLC/110/06—Order of 15 August 2007) (online). http://www.scottish-land-court.org.uk/decisions/SLC.110.06.rub.html.

Scottish Renewables. 2014. *Submission to the Smith Commission*, 3 November 2014 (online). https://www.scottishrenewables.com/publications/submission-smith-commission/.

Slee, B. 2015. Is There a Case for Community-Based Equity Participation in Scottish On-Shore Wind Energy Production? Gaps in Evidence and Research Needs. *Renewable and Sustainable Energy Reviews* 41: 540–549.

Slee, W. 2013. *A Review of Socio-Economic Considerations in Powys Wind Farm Development Proposals*. Report to Powys County Council.

Swain, N.J. 2007. Decollectivisation Politics and Rural Change in Bulgaria, Poland and the Former Czechoslovakia. *Social History* 33 (1): 1–26.

The Press and Journal. 2014. *Call to Connect Northern and Western Isles by 2020* (online). https://www.pressandjournal.co.uk/fp/news/politics/holyrood/353636/call-to-connect-northern-and-western-isles-by-2020/.

Wightman, A. 1996. *Who Owns Scotland?* Edinburgh: Canongate.

Authors' Biography

Bill Slee is an Emeritus Fellow at the James Hutton Institute. His main interests include evaluation of rural development policies, including agri-environment policy, climate change and land use, forestry and community energy. He is active on two EU Horizon 2020 projects on innovative delivery of public goods and social innovation.

Jelte Harnmeijer A systems thinker and practitioner who strongly believes in integrating research with implementation, Jelte specialises in distributed low-carbon energy solutions. Jelte is the T.B. Macaulay Renewables Fellow at the James Hutton Institute, where he focuses on how community renewables can help address pressing challenges such as food and fuel poverty, inequality, and global climate destabilisation. He is a Founding Partner at Scene Consulting, an Edinburgh-based social enterprise that assists non-specialists in meeting their renewable energy development objectives. He holds concurrent fellowships in Carbon Economics and Carbon Finance at the University of Edinburgh.

CHAPTER 4

Marine Renewables: A Distinctly Scottish Dimension

Alan Taylor

4.1 Introduction

As an established technology, onshore wind power generation has been widely deployed in the UK over the past decade. However, as planning consents and power grid capacity have become increasingly problematic for onshore wind developers to obtain, so UK renewable investments have moved towards commercial-scale offshore wind power. Although the mechanical reliability and power performance of wind turbines have now reached acceptable performance levels, wind power does have its limitations regarding the intermittency of generation, as a result of the inability to generate power in low and extreme wind conditions. Although the intermittency of wind can be partially mitigated by the wide geographic dispersal of wind farms, a country cannot rely on wind power alone to displace large amounts of its fossil-fuelled base-load generation (i.e. coal and gas). A high penetration of wind power in the energy mix of a country creates challenges in maintaining the security of supply and stable energy trading prices for wind generators (Oswald et al. 2008).

Recent advances in marine energy technologies now promise alternative renewable energy options for more consistent and predictable forms

A. Taylor (✉)
University of Dundee, Dundee, UK
e-mail: alantaylor1967@hotmail.com

G. Wood and K. Baker (eds.), *A Critical Review of Scottish Renewable and Low Carbon Energy Policy*, Energy, Climate and the Environment,
DOI 10.1007/978-3-319-56898-0_4

of generation for the UK, especially for Scotland. Despite wave power being marginally less intermittent than wind power, it is still susceptible to periods of no generation since waves are primarily driven by oceanic winds. In contrast, tidal power generation is almost completely predictable and does not suffer the intermittency problems of wind and wave. The deterministic nature of tidal power is highly desirable to power generators and network demand planners. Since the times and sizes of the tides can be calculated from the astronomical positions of the Earth, Sun and Moon, a guaranteed amount of power generation from tidal energy farms can be accurately predicted and precisely scheduled to meet network demand (Iyer et al. 2012).

The waters off Scotland's westerly coastline contain an abundance of energetic waves, driven by long-fetch Atlantic winds and amplified by subsea topography as they meet the European continental shelf. The interaction of North Atlantic and North Sea tidal systems results in high tidal levels creating powerful tidal flows in the Pentland Firth and around the Orkney coastline. The low-pressure weather conditions, frequently encountered in Scotland's northern seas, often create surge conditions complementing the normal tidal flows, resulting in even higher tides and increased tidal flows. Although past estimates of Scotland's marine energy potential have been criticised for being widely optimistic, recent hydrological studies provided more realistic estimates of the maximum electricity generation capacity potential from tidal stream of 1.9 GW (peak) in the Pentland Firth (Adcock et al. 2013) plus approximately 0.9 GW (peak) off of Islay and the Mull of Galloway (ABPmer 2007), and an additional 11 GW (mean) of wave power in the seas offshore of the Atlantic west coast and the Northern Isles (AMEC 2012).

As marine energy technology matures over the next decade, it is anticipated that marine energy generated from the oceans and seas around the world will play an increasing role in the delivery of renewable energy to fulfil national targets, with potentially significant contributions from nations such as Scotland which are naturally blessed with excellent wave and tidal resources.

4.2 Scotland's Opportunity to Take the Lead

Just as Scotland has risen to become a global technology supplier to the offshore oil and gas (O&G) industry through its exploitation of the resources beneath the North Sea, now Scotland is perfectly placed to become the global technology leader in the evolving marine energy generation industry. This potential is evident from the pioneering wave and tidal energy research being conducted by its academic establishments and the ever-growing list of indigenous start-up companies developing marine energy devices (The Institution of Mechanical Engineers 2008).

Scotland's universities and colleges can proudly boast a multitude of firmly established, world-class academic centres of excellence specialising in strategic research and technical learning areas, such as hydrodynamics, naval architecture, offshore structural engineering and marine environmental science. The academic centres in Scotland have embraced the challenge to produce a new generation of workers with the specialised skills essential for the continued growth in the fledgling marine energy industry in the next decade and beyond.

A considerable portion of the marine operational experience and offshore engineering technology from Scotland's offshore O&G sector can be easily applied into the marine energy industry, offering Scotland employment opportunities as they assert a technology lead position in this fast-evolving global industry. The International Energy Agency's (IEA) 2050 vision includes a goal of installing 337 GW of marine energy generating capacity, creating 1.2 million jobs (IEA 2013).

4.3 Scotland's Strategic Vision for Marine Energy

Although the majority of the UK energy policies affecting Scotland are still under the reserved powers of Westminster (however, see Chap. 2 in particular for a detailed explanation of the issue of reserved and devolved energy powers in Scotland and the wider UK), the Scottish Government in Holyrood has used its devolved powers wherever possible to implement a unique range of enabling policies to promote the uptake of renewable energy in Scotland, including more aggressive carbon dioxide reduction targets and ambitious targets to produce 100% of its gross electricity demand from renewable sources by 2020 (Scottish Government 2011).

The industry-led Marine Energy Group (MEG) was established in October 2003, to give guidance to the Scottish and UK Government bodies, public agencies and private sector companies on how best to fulfil Scotland's marine energy ambitions. Published in August 2009, the Marine Energy Road Map (MEG 2011) set forth a vision of how Scotland's wave and tidal energy potential could be exploited.

Subsequently, in June 2012, a summary review was issued outlining areas of progress made against the MEG action plan and highlighting where further work is still required (MEG 2012).

Scotland has delivered one of the most attractive revenue support schemes for wave and tidal stream energy in the world. In 2007, the Scottish Government, realising the opportunities for developing a marine renewables sector, established the Marine Supply Obligation to provide higher subsidies for marine renewables in comparison with the rest of the UK (Wood 2010). The Renewables Obligation (Scotland) Order 2009 (UK Government 2009a) continued this Scottish-specific focus under the Scottish version of the Renewables Obligation (the Renewables Obligation Scotland, or ROS) in April 2009, originally rewarding only two Scottish Renewable Obligation Certificates (SROCs) per megawatt-hour (MWh) for electricity generated from wave and tidal stream technologies. Although this was twice the level set for the rest of the country by the UK Government, members of the fledgling marine energy industry quickly realised that additional stimulus was required to advance the marine energy industry and petitioned the Scottish Government to provide extra revenue support for marine energy projects. Under the Scottish Statutory Instruments, the Scottish Government responded in July 2009 and the Renewables Obligation (Scotland) Amendment Order 2009 (UK Government 2009b) was enacted to provide enhanced support levels of three SROCs/MWh for tidal stream and five SROCs/MWh for wave technologies.

The generous support levels provided via the ROS have successfully attracted investment in marine energy development in Scotland from around the UK and internationally. The development cycle for tidal stream turbines has made great progress towards commercialisation with multiple full-scale prototype devices now built, deployed and tested. To further accelerate tidal stream technology, the Renewables Obligation (Scotland) Amendment Order 2013 (UK Government 2013) was introduced in April 2013 to move support levels to five SROCs/MWh for both wave and tidal stream up to a 30 MW project capacity limit, and then two SROCs/MWh above that capacity. In terms of subsidy support, and in other areas of policy, it can be seen that Scotland has consistently led the way in providing enhanced support for these early-stage, but promising technologies with the UK Government repeatedly being forced to play catch-up.

With the UK Government proposing to close the RO/ROS early, it remains to be seen whether the replacement subsidy mechanism, the Contracts for Difference Feed-in Tariff (CfD FiT), will be sufficient to continue to drive forward the development and deployment of wave and tidal stream energy. Of particular relevance to the Scottish marine renewables sector, the CfD FiT has returned control over subsidy levels to Westminster with many arguing that projects will not be 'bankable' due to the lack of clarity regarding support for renewables under the new CfD FiT regime post-2020.

4.4 Streamlining of Marine Planning and Consenting Processes

Holyrood has also embarked on a mission to implement progressive measures aimed at streamlining the licensing and planning consent processes for renewable energy schemes (Scottish Government 2012). In the past 4 years, reforms under the Marine (Scotland) Act 2010 (Scottish Government 2010) have drastically overhauled and streamlined the consenting process for marine energy developers via the creation of a new '*one-stop shop*' marine management organisation for issuing licenses and consents for deployment of wave and tidal devices, the Marine Scotland Licensing Operations Team.

In March 2007, a Strategic Environmental Assessment was announced to examine the environmental effects of developing wave and tidal schemes (Scottish Government 2007). To further improve the Scottish marine planning process, the Environmental Impact Assessment (EIA) planning legislation was revised and implemented in June 2011 to encompass marine planning aspects for offshore renewable energy schemes (UK Government 2011). The Scottish Marine Research Unit (SMRU) and Project Management Support Services Limited (PMSS)[1] have been commissioned as consultants by the UK Government research body, the Natural Environment Research Council (NERC), RenewableUK and Scottish Renewables to produce a series of papers consolidating existing knowledge to give additional guidance for wave and tidal energy consenting, specific to environmental impacts on ornithology, marine mammals, fish and shellfish (SMRU Consulting 2013).

4.5 Technology Maturity of Marine Energy Machines

The offshore marine environment can be a hostile place in which wave and tidal energy machine must operate reliably. Consequently, progressive development and demonstration steps are required before a fully developed energy system can be sold commercially. It is essential to prove reliable operation, energy output levels and economic viability to minimise risks for developers of marine energy schemes.

The majority of today's wave and tidal energy converters do not have sufficient operational experience to long term demonstrate their reliability, but do face many of the same environmental challenges as offshore oil and gas (O&G) installations. In May 2005, the international marine certification body Det Norske Veritas (DNV) in conjunction with the Carbon Trust adapted their extensive experience in the O&G industry to produce guidelines on the design and operation of wave energy converter (WEC) devices based on DNV's comprehensive offshore design standards and recommended practices (DNV-Carbon Trust 2005).

The DNV-Carbon Trust guideline document was a precursor to the first offshore standard for the certification of tidal and wave energy converters which was published in October 2008. The DNV-OSS-312 standard (Certification of Tidal and Wave Energy Converters) aimed to provide an independent verification of acceptable performance (safety, availability, reliability, asset integrity and environmental impact) to financiers, investors, utility companies, insurers and the public (DNV 2012).

The maturity of marine energy technologies can be assessed using a 9-point ranking scale indicating its Technology Readiness Level (TRL); this is an adaption of a ranking system first devised for the space industry (NASA 2010). The TRL ranking scale provides a consistent and easy-to-understand system for marine energy developers and investors to evaluate technology and commercial risks, from the early conceptual stages through to proven full-scale commercial machines (see Table 4.1).

The Ocean Energy System (OES) group of the International Energy Association has further simplified the TRL ranking system into five discrete stages for development, with each stage providing various aspects of guidance for system design, simulation, testing and economic assessment (IEA-OES 2010). This simplified methodology is becoming a '*de facto*' protocol for developers and is promoted by many marine energy testing facilities including EMEC (see Table 4.2). The OES is assisting in the development of a full suite of international standards for marine energy.

Table 4.1 Technology readiness levels for marine energy technologies

TRL	*Description*	*Scale*	*Outputs*
1	Configuration described	n/a	Basic principles observed and reported
2	Technology stream initiated	n/a	Scale of technology concept formulated
3	Initial product verification	>1:100	Analytical and experimental critical function and/ or characteristic proof-of concept
4	Laboratory and analytical verification	>1:25	Technology component and/or basic technology subsystem validation in a laboratory environment
5	Reduced-risk subsystem verification	>1:15	Technology component and/or basic technology subsystem validation in a relevant environment
6	Reduced-risk full system verification	>1:4	Technology system prototype demonstration in a relevant environment
7	Ocean operational readiness	>1:2	Technology system prototype demonstration in an operational environment
8	Pre-commercial project readiness	1:1	Actual product (first of type) completed and qualified through test and demonstration
9	Commercial project readiness	1:1	Operational performance and reliability demonstrated for an array of type machines

Source Fitzgerald and Bolund 2012

Table 4.2 Recommended development protocols for wave and tidal energy

Stage	*TRL*	*Wave energy protocol*	*Tidal energy protocol*
1st	1, 2, 3	Concept validation. Prove the basic concept from wave flume tests in small scale	Tidal-current energy conversion concept formulated
2nd	4	Design validation. Subsystem testing at Intermediate scale, Flume tests scale 1:10, Survivability; Computational fluid dynamics; Finite element analysis Dynamic analysis, engineering design (Prototype); feasibility and costing	Intermediate scale subsystem testing, Computational fluid dynamics, Finite element analysis, Dynamic analysis
3rd	5, 6	Subsystem testing (full-scale). Testing operational scaled models at sea, and subsystem testing at large scale	Subsystem testing at large scale
4th	7, 8	Prototype Testing (full-scale). Full-scale prototype tested at sea	Full-scale prototype tested at sea
5th	9	Economic validation. Several units of precommercial machines tested at sea for an extended period of time	Commercial demonstrator tested at sea for an extended period

Source IEA-OES 2010

4.6 Moving from Concept to Commercialisation

The combination of good natural resources, a highly skilled workforce, a simplified marine planning framework and a supportive economic environment, has seen significant investments in Scotland to develop the first commercial marine energy arrays and farms in its territorial waters (Carbon Trust 2013). As the first commercial arrays are deployed, the primary challenge facing the marine energy industry is to reduce the Cost of Energy (CoE) by the mid-2020s to a point where it is competitive with other renewable technologies. Recent estimates by the Carbon Trust have put the Levelised Cost of Energy (LCoE) for tidal energy in the range £290–330 per MWh and £380–480 per MWh for wave energy.

The Carbon Trust has identified clear opportunities for the reduction in the CoE for marine energy generation, placing a focus on supporting technology innovation through its Marine Energy Accelerator (MEA) programme. The experiences gathered from deployment and operation of the first arrays will provide valuable learning to drive the technology innovation bringing down the capital and operational costs of marine energy to give a LCoE of £150–200 per MWh for wave and tidal technologies by 2020 (Carbon Trust 2011).

4.7 Supporting Innovation in Marine Energy

Due to the high costs, funding assistance for innovation and demonstration of marine energy technologies is often essential for small and medium-sized enterprises, the typical scale of company developing marine technology particularly at the early stage of prototype device research, design and testing, to prove their prototypes, and is considered as a key enabler for accelerating the deployment rate of marine energy in Scotland and the wider UK. In 2003, the Carbon Trust launched the '*Marine Energy Challenge*' to develop understanding and overcome barriers facing the marine energy industry as it progressed towards commercialisation.

The Carbon Trust is one of the UK's leading independent authorities on marine energy, managing multiple funding programmes focussed on reducing the cost of energy for commercial-scale marine energy deployments including the £22.5 m Marine Renewables Proving Fund (MRPF), and the Scottish Government's £18 m Marine Renewables

Commercialisation Fund (MRCF). The Carbon Trust, in collaboration with the offshore renewable energy (ORE) Catapult centre recently launched the Marine Farm Accelerator (MFA) programme for the development and proving of essential technology for the first marine energy arrays.

Other joint private–public funding via the Technology Strategy Board (TSB) and Energy Technologies Institute (ETI), a private–public partnership to promote and manage the delivery of key energy technology programmes, has provided assistance for technology development in marine energy, such as the £10.5 m Marine Energy Supporting Array Technologies (MESAT) competition launched in March 2012 with Scottish Enterprise and NERC.

The capital costs of the first demonstrator arrays of wave and tidal devices can be considerable, and financial assistance has been provided by both the UK and Scottish Governments. In February 2013, DECC awarded £20 million of funding to develop wave and tidal technologies from prototype stage to the demonstration of arrays of devices through its Marine Energy Array Demonstrator (MEAD) initiative; this included part funding of the construction costs of Meygen's tidal turbine array in the Inner Sound of the Pentland Firth.

The Scottish Government, in collaboration with Scottish Enterprise and HIE, secured funding from the European Regional Development Fund (ERDF) to provide £6 m for the development of new wave and tidal stream prototypes and related infrastructure via their Wave and Tidal Energy: Research, Development and Demonstration Support (WATERS1) fund. In August 2012, the Scottish Government followed this by the second funding round (WATERS2), releasing further grants totalling £7.9 m to marine energy developers in Scotland, including £1.2 m support for Scotrenewables to develop their 2 MW commercial-scale tidal turbine.

4.8 Academic Research and Skills Training for Marine Energy Sector

In October 2003, the Engineering and Physical Sciences Research Council (EPSRC) funded the creation of the SuperGen programme, a flagship research initiative to shape the future of the UK's energy research landscape. The first phase of the SuperGen programme brought together an academic research consortium for marine energy comprising of the Universities of Edinburgh, Strathclyde, Robert Gordon, Heriot-Watt and Lancaster with the focus on developing the learning to exploit the potential of marine energy resources. SuperGen Marine Phase 1 (2003–2007) produced valuable research in marine energy and trained 13 Ph.D. graduates who subsequently took up employment in influential positions across the marine energy sector.

Building on the success of Phase 1, SuperGen Phase 2 (2007–2011) saw the marine consortium expanded, bringing together research staff from Universities of Edinburgh, Queen's Belfast Heriot-Watt, Lancaster and Strathclyde at the core, including affiliate Universities of Durham, Exeter, Highlands and Islands, Manchester, Robert Gordon and Southampton. Phase 2 was advised by an Industry Research Advisory Forum including Marine Current Turbines, Open Hydro (DCNS), Pelamis Wave Power (PWP), Ocean Power Technology (OPT), Scottish & Southern Energy (SSE), Scottish Power Renewables (SPR), EdF, e-On, nPower, EMEC, NaREC, The Crown Estate and Scottish Natural Heritage (SNH). Phase 2 also saw collaboration with many international partners, including: HMRC Cork (Ireland), TU Delft-(Netherlands), École Centrale Nautique (Nantes, France), Dalhousie University (Canada), Oregon State University, Florida Atlantic University, UMass (USA), Universities of Osaka City and Hokkaido (Japan), Harbin Engineering University and Dalian University of Technology (China), National Sun Yat-Sen, National Taiwan Ocean University and National Chen Kung University (Taiwan). Over 30 Ph.D. graduates were funded during Phase 2 to supply the marine energy sector with the high-level skills.

The SuperGen Marine programme secured finance for Phase 3 to fund five further research years from 2011. Phase 3 brings together the Universities of Edinburgh, Queen's Belfast, Strathclyde and Exeter as core research establishments with Universities of Plymouth, Heriot-Watt, Lancaster, Manchester, Swansea, Oxford and Southampton as associates. Together, the Universities form the UK Centre for Marine Energy Research to provide joined-up regional, disciplinary and thematic research to meet the challenges of accelerating marine energy deployment.

The Energy Technology Partnership (ETP) is a consortium of Scottish universities formed to develop energy technologies in conjunction with industrial partners. Marine energy is one of the key themes supported by the ETP. The inter-university alliance pools both resources and expert knowledge across a multidisciplinary spectrum, including planning policy, environmental impacts, resource assessment, economics modelling, fluid dynamics, engineering design and prototype testing. The ETP universities have some world-class marine test facilities at their disposal. A marine energy sector partnership with the ETP universities has brought together marine test facilities under the Scottish Energy Laboratory umbrella. Facilities of particular relevance to the marine sector include:

- FloWave TT—All Waters Combined Current and Wave Test Facility (University of Edinburgh)
- Curved wave tank and wave flumes (University of Edinburgh)
- Kelvin Hydrodynamics Laboratory (University of Strathclyde)
- OceanLab sea testing facility (University of Aberdeen)
- Heriot-Watt University wave basin
- Machine and power electronics test laboratory (University of Edinburgh)
- Energy Technology Centre—component test facilities
- European Marine Energy Centre (EMEC).

4.9 Proving of Commercial-Scale Marine Energy Prototypes

Recognising the development potential of Orkney's marine energy resources, the Scottish Government jointly funded the creation of EMEC in 2003, to provide the world's first purpose-built site for prototype testing of wave and tidal energy converters. The accredited marine test centre provides developers with pre-consented, open-sea testing facilities for their machines, complete with subsea electrical connection points to the onshore power network.

After a decade of successful operation, EMEC has now established itself as the world's leading centre for the testing and proving of marine energy converters. Developers from around the globe have benefitted from EMEC's wave test berths at Billa Croo and tidal test berths at the Fall of Warness, to advance their marine energy devices towards commercial viability (see Tables 4.3 and 4.4).

Table 4.3 Full-scale testing of wave energy converters in Scotland

Wave Energy Developer	*Test site*	*WEC device*	*Capacity (kW)*
Aquamarine power	Billia Croo, EMEC	Oyster 800	800
Pelamis wave power	Billia Croo, EMEC	Pelamis P2	750
Seatricity	Billia Croo, EMEC	Oceanus	800
Wello Oy	Billia Croo, EMEC	Penguin	500

Source EMEC 2015

Table 4.4 Full-scale testing of tidal energy converters in Scotland

Tidal energy developer	*Test site*	*Tidal turbine device*	*Capacity*
Andritz Hydro Hammerfest	Fall of warness, EMEC	HS1000	1 MW
OpenHydro (DNCS)	Fall of warness, EMEC	Open centre turbine	250 kW
Alstom (formerly TGL)	Fall of warness, EMEC	TFL DeepGen	1 MW
Voith Hydro	Fall of warness, EMEC	HyTide 1000-13	1 MW
Bluewater energy services	Fall of warness, EMEC	BlueTEC	2 MW
Atlantis resources corp	Fall of warness, EMEC	AR1000	1 MW
Scotrenewables	Lashy sound, EMEC	SRI 500	1.5 MW
Kawasaki heavy industries	Fall of warness, EMEC	Kawaski	1 MW
Nautricity	Shapinsay sound, EMEC	CoRMaT	500 kW

Source EMEC 2015

4.10 Challenge to Scale-up—The Saltire Prize

As an extra stimulus to innovate and accelerate marine energy machines towards full-scale commercialisation, the Saltire Prize was announced by the Scottish Government. A £10 m prize was offered to competition entrants who generated the most electricity from wave or tidal energy over a two-year continuous period, with a minimum generation output of 100 GWh.

The Saltire Prize competition has now entered the *'Grand Challenge'* phase with four entrants declared at the entry deadline of January 2015, following the financial collapse of Pelamis Wave Power in late December 2014. The marine energy projects include one wave and three tidal energy schemes:

- North-west Lewis Wave Power Project (Aquamarine Power), 40 MW wave array.
- MeyGen Tidal Project, Phase 1 (Atlantis Resources Corp.), 86 MW tidal array.

- Ness of Duncansby Tidal Project (Scottish Power Renewables), 30 MW tidal array.
- West Islay Tidal Project (DP Marine Energy), 30 MW tidal array.

4.11 Conclusion

Over the past decade, the Scottish Government has consistently provided financial support as a stimulus for technology innovation in the marine energy industry sector. Significant hurdles in the technology development cycle have already been crossed, but now the main challenge is to prove commercial viability. The challenge in engineering and manufacturing cost-competitive wave and tidal machines is not trivial, due to the high-quality specifications required to meet the desired levels of operational reliability and survivability in the aggressive ocean environment.

This nascent sector is at last showing signs of technology maturity, with the deployment of the first commercial full-scale wave and tidal scale arrays planned in Scottish waters in the next few years, creating a fresh spirit of optimism in the emergent marine energy industry throughout the wider UK and globally. Indeed, the first tidal turbine of the five-turbine Shetland Tidal Array is already delivering power to the Shetland grid, with another four turbines planned (Renews 2016). Major international energy companies are now investing in marine energy technology companies, through stake-holding purchases and acquisitions. The entry of these big players has significantly reduced the risk of commercial failure for many small/medium-sized wave and tidal energy device developers.

Considering Scotland's existing technical skill base and its unrivalled innovative research in marine energy, the challenges in reaching full commercialisation are certainly not insurmountable. Scotland already possesses world-leading offshore expertise from its strong engineering background in shipbuilding and offshore oil. The UK Government and many other national governments in North-western Europe, Asia and North America have openly acknowledged Scotland's pioneering endeavours to advance the challenging marine energy sector, whilst also recognising the high costs in developing the technologies to demonstrate the prerequisite levels of performance and reliability for attracting commercial energy investors. It is almost certainly outwith the ability of one small government (Scottish) to 'go-it-alone' and future intergovernmental collaborations are vital to provide the levels of financial assistance to enable marine energy technology developers overcome the challenging technology barriers on the route to proving commercial viability.

Scotland is uniquely placed to establish itself as the world-leading centre of marine energy technology, resulting in major employment and global export opportunities for the supply of goods and services for wave and tidal schemes throughout the world. Greater challenges were successfully overcome in the pursuit of North Sea oil and gas over the past 40 years, producing considerable returns to the UK economy. Establishing a similar technology leadership in the marine energy sector would result in substantial benefits for the Scottish nation, as the world moves away from its dependency on hydrocarbons towards renewable energy sources.

Note

1. PMSS provides consultancy exclusively to renewable energy sector, now part of TÜV SÜD, see http://www.pmss.com.

References

ABPmer. 2007. *Quantification of Exploitable Tidal Energy Resources in UK Waters: A Report Commissioned by nPower Juice, Report No. R1349*. July 2007, vol. 1. ABP marine environment research.

Adcock, T.A.A., S. Draper, G.T. Houlsby, A.G.L. Borthwick, and S. Serhadlioglu. 2013. The Available Power from Tidal Stream Turbines in the Pentland Firth. In *Proceedings of the Royal Society*, Sept 8, 2013, vol. 469, no. 2157 p. 20130072.

AMEC. 2012. Carbon Trust UK wave energy resource. October 2012, Carbon Trust, UK [Online]. http://www.carbontrust.com/media/202649/ctc816-uk-wave-energy-resource.pdf. Accessed 29 Sept 2015.

Carbon Trust. 2011. Accelerating marine energy: The potential for cost reduction - insights from the Carbon Trust Marine Energy Accelerator. July 2011, Carbon Trust, UK [Online]. http://www.carbontrust.com/media/5675/ctc797.pdf. Accessed 29 Sept 2015.

Carbon Trust. 2013. Is marine energy about to go mainstream? February 2013, Carbon Trust, UK [Online]. http://www.carbontrust.com/news/2013/02/is-marine-energy-about-to-go-mainstream. Accessed 29 Sept 2015.

DNV-Carbon Trust. 2005. Guidelines on design and operation of wave energy converters. May 2005, DNV, Norway [Online]. http://www.gl-group.com/pdf/WECguideline_tcm4-270406.pdf. Accessed 29 Sept 2015.

DNV. 2012. DNV-OSS-312: Certification of Tidal and Wave Energy Converters. April, 2012, Det Norske Veritas AS (DNV), Norway [Online]. https://exchange.dnv.com/publishing/Codes/download.asp?url=2012-04/oss-312.pdf. Accessed 29 Sept 2015.

European Marine Energy Centre. 2015. Our Sites [Online]. http://www.emec.org.uk/about-us/our-sites/. Accessed 29 Sept 2015.

Fitzgerald, J., and B. Bolund. 2012. Technology Readiness for Wave Energy Projects; ESB and Vattenfall Classification System. ICOE 2012, Dublin, Ireland [Online]. http://www.icoe-conference.com/publication/technology_readiness_for_wave_energy_projects_esb_and_vattenfall_classification_system/. Accessed 29 Sept 2015.

International Energy Association (IEA). 2013. Energy Technology Initiatives: Implementation Through Multilateral Co-operation 2013. IEA, December 2013 [Online]. http://www.iea.org/publications/freepublications/publication/EnergyTechnologyInitiatives_2013.pdf. Accessed 29 Sept 2015.

International Energy Association Ocean Energy Systems (IEA-OES). 2010. Development of Recommended Practices for Testing and Evaluating Ocean Energy Systems (Annex II Extension): Summary Report, OES IA Document No: T02-00. August, 2010 [Online]. http://www.ocean-energy-systems.org/oes_reports/annex_ii_reports/development_of_recommended_practices_for_testing_and_evaluating_ocean_energy_systems_2010_summary/. Accessed 29 Sept 2015.

Iyer, A.S., S.J. Couch, G.P. Harrison, and A.R. Wallace. 2012. Variability and Phasing of Tidal Current Energy Around the United Kingdom. *Renewable Energy* 51 (2013): 343–357.

Marine Energy Group (MEG). 2011. FREDS Marine Energy Group (MEG): Marine Energy Road Map. Scottish Government, August 2011 [Online]. http://www.scotland.gov.uk/Resource/Doc/281865/0085187.pdf. Accessed 29 Sept 2015.

Marine Energy Group (MEG). 2012. Marine Energy Group: Marine Energy Action Plan. Scottish Government, June 2012 [Online]. http://www.scotland.gov.uk/Resource/0039/00395516.pdf. Accessed 29 Sept 2015.

NASA. 2010. Technology Readiness Level Definitions. May 2010, NASA, USA [Online]. http://www.nasa.gov/pdf/458490main_TRL_Definitions.pdf. Accessed 29 Sept 2015.

Oswald, J., M. Raine, and H. Ashraf-Ball. 2008. Will British Weather Provide Reliable Electricity? *Energy Policy* 36 (8): 3212–3225.

Renews. 2016. First Power at Shetland Tidal Array [Online]. http://renews.biz/101854/first-power-at-shetland-tidal-array/. Accessed 26 Oct 2016.

Scottish Government. 2007. Scottish Marine Renewables: Strategic Environmental Assessment (SEA) [Online]. http://www.scotland.gov.uk/publications/2007/03/seawave. Accessed 29 Sept 2015.

Scottish Government. 2010. Marine (Scotland) Act 2010 (asp 5) [Online]. http://www.legislation.gov.uk/asp/2010/5/pdfs/asp_20100005_en.pdf. Accessed 29 Sept 2015.

Scottish Government. 2011. 2020 Routemap for Renewable Energy in Scotland. July 2011, Scottish Government, 2011 [Online]. http://www.scotland.gov.uk/Resource/Doc/917/0118802.pdf. Accessed 29 Sept 2015.

Scottish Government. 2012. Scotland's Renewables Routemap: Short Life Task Force on Streamlining Energy Development Licensing and Consents—Final Report [Online]. http://www.scotland.gov.uk/resource/0038/00387242.pdf. Accessed 29 Sept 2015.

SMRU Consulting. 2013. Wave and Tidal Energy Review. News article—September 2013, SMRU Consulting [Online]. http://www.smruconsulting.com/wave-and-tidal-energy-review/. Accessed 29 Sept 2015.

The Institution of Mechanical Engineers. 2008. Marine Energy; More Than Just a Drop in the Ocean. December, 2008, Institute of Mechanical Engineers, UK [Online]. http://www.imeche.org/knowledge/themes/energy/energy-supply/renewable-energy/marine-energy. Accessed 29 Sept 2015.

UK Government. 2009a. The Renewables Obligation (Scotland) Order 2009. Scottish Statutory Instruments 2009, No. 140, Electricity. March 2009 [Online]. http://www.legislation.gov.uk/ssi/2009/140/pdfs/ssi_20090140_en.pdf. Accessed 29 Sept 2015.

UK Government. 2009b. The Renewables Obligation (Scotland) Amendment Order 2009. Scottish Statutory Instruments 2009, No. 276, Electricity. July 2009 [Online]. http://www.legislation.gov.uk/ssi/2009/276/pdfs/ssi_20090276_en.pdf. Accessed 29 Sept 2015.

UK Government. 2011. The Town and Country Planning (Environmental Impact Assessment) (Scotland) Regulations 2011 [Online]. http://www.legislation.gov.uk/ssi/2011/139/pdfs/ssi_20110139_en.pdf. Accessed 29 Sept 2015.

UK Government. 2013. The Renewables Obligation (Scotland) Amendment Order 2013. Scottish Statutory Instruments 2013, No. 116, Electricity. March 2013 [Online]. http://www.legislation.gov.uk/ssi/2013/116/pdfs/ssi_20130116_en.pdf. Accessed 29 Sept 2015.

Wood, G. 2010. *Renewable Energy Policy in Scotland: An Analysis of the Impact of Internal and External Failures on Renewable Energy Deployment Targets to 2020*. CEPMLP, Dundee: CEPMLP Energy Series.

Author Biography

Alan Taylor is a chartered electrical engineer, with over 25 years' engineering experience of high reliability electrical systems for use in harsh environments. Currently, he is working on the completion of his Ph.D. focused on the engineering asset management aspects of offshore renewable energy farms.

CHAPTER 5

Renewable Heat: The Perfect Storm?

Keith Baker

5.1 Introduction

A perfect storm is brewing. No single area of policy has a greater potential to undermine the Scottish Government's energy and climate change targets than the provision of renewable heat, and if it fails, then those hit hardest will be the fuel poor, and particularly those in the islands and rural areas.

When the Scottish Government set itself the target of meeting 100% of its electricity demand from renewables by 2020 (Scottish Government 2011a), it already knew that the greatest challenge would be sustaining that target, as more and more homes and businesses switch to electricity and low-carbon alternatives for heating (primarily biomass and Combined Heat and Power (CHP) systems); and it already knew that its poorly maintained building stock and the expected rises in gas prices could drive the fuel poor further into fuel poverty (Baker et al. 2012a). So when in late 2013 the energy companies announced above-inflation price rises and warned of worse to come, the only people who could claim to be surprised were the public and those in the media who had been hoodwinked into believing that the answers lay merely in better regulation of the energy industry.

This is not to say that better regulation is not needed, but that it is only part of the solution to a problem for which the fundamental cause

K. Baker (✉)
School of Engineering and the Built Environment,
Glasgow Caledonian University, Glasgow, UK
e-mail: keith.baker@gcu.ac.uk

G. Wood and K. Baker (eds.), *A Critical Review of Scottish Renewable and Low Carbon Energy Policy*, Energy, Climate and the Environment,
DOI 10.1007/978-3-319-56898-0_5

is short-term political thinking that has been decoupled from headline-grabbing long-term targets. This chapter will explore what went wrong and what options exist for pulling Scotland back from the post-2020 precipice.

5.2 What Went Wrong? Political Footballs

Let us first deal with the role of the energy companies. Whilst much is made of the 'Big Six' suppliers raising energy prices and lobbying against levies such as the Green Deal and Energy Company Obligation (ECO), it is easy to forget that the UK still pays some of the lowest energy prices in Europe. For example, in May 2013, the cost of natural gas in the UK was just under €0.05 per kWh, whereas in Denmark and Sweden, which operate much more enlightened approaches to heat policy, the unit prices were more than double (EEP 2013). Yet, despite this, Denmark has significantly and consistently lower levels of fuel poverty, whilst the home of the Enlightenment fares even worse than the rest of the UK (Snodin 2008).

Other criticisms commonly levelled at the energy companies relate to their failures in delivering schemes such as the Carbon Emissions Reduction Target (CERT), the Community Energy Saving Programme (CESP), the Green Deal, ECO and (in Scotland) the Energy Assistance Package (EAP) (Baker et al. 2012a, 2014; DECC 2013). However, on closer inspection it becomes clear that the failures of these schemes, in particular the disastrous Green Deal and ECO, relate as much to their design as their implementation (see Chapter 7). Here again, this is not to say the energy companies should be absolved of responsibility for these failures—indeed quite the opposite given their lobbying for the removal of the levies needed to fund them from household energy bills—but to point to the responsibilities of both the Westminster and Holyrood Governments for not taking on board years of clear and (relatively) consistent advice from independent experts.

A common and fundamental criticism, and no doubt a valid one, is that the industry should never have been privatised, but at least until the Scottish National Party's much enhanced contingent at Westminster are seen to be lobbying hard for renationalisation, the Scottish Government cannot claim to want to make the more significant changes to the industry that many experts would advocate. Opinion polls and campaign groups consistently reflect public outrage over executive salaries, but governments consistently refuse to rein them in, whilst the Department for Energy and Climate Change seemed positively enthralled by any crumbs it got fed by the industry, and Holyrood seemed powerless to intervene. To make matters even worse, in the summer of 2015, Chancellor of the Exchequer George Osborne quietly moved much of the responsibilities for energy policy into the Treasury (HM Treasury 2015), from where he then began the process of dismantling it completely, leaving DECC with little to do but wave the flag for fracking and manage the on-going train wreck that is the government's push for new nuclear. And it did not even end there, as one of Prime Minister Theresa May's first actions was to scrap DECC entirely.

The Scottish Government's historic fallback position of blaming Westminster for the failures of non-devolved policies and programmes would of course have been removed under independence, and following England and Wales voting to leave the European Union a second referendum before 2020 is now very much back on the cards, but the degrees to which the failings of energy efficiency and fuel poverty schemes can be attributed to Westminster are highly questionable. Whilst there is good evidence to show that devolved programmes such as the EAP and the Home Insulation Scheme (HIS) have proved more effective than their non-devolved counterparts (Maiden et al. 2016), it should be noted that the design and funding of these schemes has been built on the assumption that Scotland would be able to leverage the maximum benefits of the UK-wide programmes (Baker et al. 2012a, b; Scottish Government 2010, 2011b). Therefore, in the light of expert advice and the often-voiced political scepticism of the effectiveness and appropriateness of these non-devolved schemes, it would seem incredibly foolish not to base devolved policies on anything other than worst-case scenarios, and indeed two of those worst-case scenarios have now been realised in the form of the failures of the Green Deal and ECO. This leads us to a problem closely interlinked with the provision of renewable heat, that of fuel poverty.

5.3 What Went Wrong?—Fuel Poverty

Fuel poverty is an increasingly contentious subject and a growing source of embarrassment for the Scottish Government, which back in 2002 (as the Scottish Executive) set itself the target of eliminating the problem by November 2016 (Scottish Executive 2002). Yet, despite this lofty ambition the number of Scottish households in fuel poverty rose sharply from 13% in 2002 to 27% in 2008, and aside from a peak of 33% in 2009 attributable to the recession, the general trend has been one of a gradual increase. And then, in 2014, the Scottish House Conditions Survey (SHCS) figures were updated using a new version of the Building Research Establishment's Domestic Energy Model (BREDEM), which revised those figures upwards and put the most figure (for 2013) at 39.1%, the highest on record (Scottish Government 2014).

In Scotland, a household is classified as being in fuel poverty if, in order to maintain a satisfactory heating regime, they would be required to spend more than 10% of their household income (including housing benefit or income support) on all household heating fuel use. The heating regime is consistent with the guidance from the World Health Organisation (WHO 1990) and adjusted for elderly and disabled occupants. This was consistent with the rest of the UK until the Westminster Government adopted the low-income high-costs 'Hills Definition' for England in 2013 (Hills 2012) and simultaneously slashed its fuel poverty statistics. At the time of writing, the Scottish Government is committed to maintaining the 10% definition; however, with the failure to meet the 2016 target, it may be difficult for politicians to resist the temptation, and for civil servants to resist the political pressure, to mitigate some of the damage by adjusting it without sufficient scrutiny. And whilst it must be acknowledged that the current definition captures a small number of high-income households in large homes who could reasonably be expected to pay for their high-energy use, amending the definition to remove these has been estimated to reduce the figure by 3–4% (Restrick 2013).

In addition, the urban–rural divide means cost of energy to Scottish households also varies significantly across the country due to levels of connection to the gas grid and the continued prevalence of using oil and other non-electric forms of heating in rural areas and the islands. At the extreme end of this scale is Eilean Siar (the Western Isles) where over 50% of households, and 83% of pensioner households, now live in fuel poverty (Scottish Government 2013a). And furthermore, recent research using actual (as opposed to estimated or modelled) household energy data shows that the additional energy spend attributable to living in rural and island communities is more significant than current statistics suggest (Mould et al. 2014).

Nevertheless, and regardless of what the future figures may be, one has to ask why fuel poverty levels have continued to rise in the first place. Obviously one part of that answer lies in the rising price of fuel, but if that were the sole significant factor one would expect a good correlation between fuel poverty and overall poverty, as measured by the Scottish Indices of Multiple Deprivation (Scottish Government 2013b); however, plotting the data from these two metrics shows that this is simply not the case (Mould et al. 2014). What does appear to be happening is that Scottish energy efficiency and fuel poverty programmes are insulating households from the impacts of energy price fluctuations, but not against fuel poverty per se (Maiden et al. 2016). Another part of the answer lies in the assumptions used in predicting energy demand for heating.

5.4 What Went Wrong?—Believing the Models

Around 75% of Scotland's heating demand is from buildings, with industrial processes making up the remainder (Scottish Government 2009). The main models used for calculating energy demand are the Standard Assessment Procedure (SAP) and its simplified variant the Reduced Data SAP (rdSAP) for domestic buildings[1] and the Simplified Building Energy Model (SBEM) for non-domestics. However, both of these tools have been widely criticised, not least for under-estimating demand and particularly for being inappropriate for use in Scotland, with its more northern climate and distinctly different building stock (for example: Baker et al. 2012a, b; Beckmann and Roaf 2013; CFS 2012; Jones Lang LaSalle 2012; Sanders and Phillipson 2006; UKGBC 2010). Indeed, SAP is so flawed that a report on social housing by Affinity Sutton (Affinity Sutton 2013) found that it overpredicts energy efficiency savings by an astounding average of 77%. At an aggregate level, the Scottish Government also uses its Domestic Energy Model for Scotland (DEMScot2), but this has also been criticised for likely under-predicting demand (CAR 2009), but if the Scottish Government's recent experience with using BREDEM for calculating fuel poverty figures has not finally instilled in it a healthy distrust of modelling, it is difficult to imagine what would. And furthermore, it is not just the demand models that have proven to be suspect, but also those used for modelling generation (Institute for Mechanical Engineers 2011).

Yet, despite this growing wealth of evidence and advice to the contrary, not least the access it has to very real figures for measured energy generation and consumption, the Scottish Government has persisted in its belief that the figures generated by the most commonly used models are fit for use in evidence-based policy-making and long-term planning.

5.5 What's Wrong with the Models?

The simple answer here is that there is still insufficient measured data available, or at least without restrictions and available in one place, from which to revise assumptions and adjust national projections. The most effective solution, and one which the Scottish Government is uniquely positioned to implement, is to collect that data from existing buildings and ensure all new build is subject to post-occupancy evaluation. It is, at last, now developing policies and proposals, such as the National Heat Map, that are moving in that direction, but these still lack sufficient granularity to be useful for local and individual decision-making. Yet experts have been making the case for a centralised database of individual building energy use for well over a decade (Bruhns et al. 2000).

On an individual dwelling basis, one of the main reasons that the models can be wildly inaccurate is that they assume a standard pattern of occupant behaviour, and this is often not replicated in practise. For example, the number of people working one or more days a week from home has increased significantly since the turn of the century, and even relatively cursory studies have found this to be a statistically significant factor in influencing energy consumption. And there is good evidence from which to question the 'zone 1/zone 2' convention that living rooms are heated more and for longer (Baker and Rylatt 2008), yet the standard assumptions persist in spite of the evidence.

Of course, occupancy behaviour can be highly varied and complex, so only when the models are scaled up to a reasonable number of properties should the models be expected to reflect actual consumption, but scaling up can also magnify any errors in the assumptions. Tenure, household size and occupant demographics are also important factors that influence the accuracy of models, and as these change, we would expect an increasing divergence between predictions and measurements—and this is exactly what appears to be happening. Furthermore, one of the main benefits of models to policy makers is to allow them to assess the costs and benefits of funding different energy efficiency measures, but this requires assumptions to be made about the building stock, and this is an area where there is a highly problematic degree of uncertainty.

An example of this is a meta study conducted in 2006 for the Energy Saving Trust, which focussed on energy consumption before and after loft insulation and cavity wall insulation measures had been installed (Sanders and Phillipson 2006). It found that that 15% of the potential savings were 'taken back' by the consumers as improved thermal comfort, i.e. people took the opportunity of improved thermal performance to achieve better internal conditions after refurbishment. In addition, the study identified that typically a further 35% of the potential savings were not realised, which was attributed to a combination of:

- Actual insulation performance once installed being less than in an ideal application. This could in part be attributed to workmanship issues, but also unrealistic assumptions about the building stock and the accessibility of particular cases for good installation of cavity insulation completely throughout the building.
- Occupant-controlled ventilation behaviour. There is anecdotal evidence to suggest that some occupants post-refurbishment have a tendency to control internal comfort when the house heats more quickly than previously, by opening windows and dumping excess heat rather than changing their thermostat settings. This behaviour might evolve with time post-refurbishment.
- Limitations in the theoretical models and assessment approaches adopted by the different studies.

In total, the EST study identified that the average actual energy saving after cavity wall insulation and loft insulation is 50% less than could theoretically be achieved. Similar behaviours have been found for other energy efficiency measures, a phenomenon known as Jevon's Paradox or the 'rebound effect', and a related effect, the prebound effect has also now been identified as a likely contributing factor (Maiden et al. 2016). Scotland's bespoke and comparatively sophisticated aggregate model, DEMScot2, remains one of the few energy models to include an adjustment for the rebound effect, but whilst this simple assumption may be justifiable in an aggregate model, dwelling-level models should be able to incorporate much more sophisticated adjustments.

In addition to the data loggers that are the staple equipment for many building scientists, smart meters are providing an increasingly powerful means of collecting data on energy consumption; however, they are not without their detractors, and for some non-domestic buildings the data is classed as commercially confidential. The UK-wide roll-out to domestic properties and small businesses has a completion target of 2020, but the models being distributed offer very limited functionality and so, whilst only a small minority of households closely monitor their energy consumption (Scottish Government 2014), the exercise looks increasingly like a missed opportunity to normalise their use (see Chapter 7). However, whilst smart meters can remotely record consumption, and by that enable us to make better inferences about occupant behaviour, they can tell us little about the condition of the building stock—which directly influences heating demand and the economic and environmental returns on energy efficiency interventions. And, especially following the cancellation of the German smart metering programme, it is arguable that they are yet another distraction from this fundamental problem.

5.6 What Went Wrong?—Maintenance, Monitoring and Assessment

According to the Scottish Household Condition Survey (Scottish Government 2014), almost half of the Scottish homes are classified as being in need of urgent repair, and even relatively small amounts of damage can have a measurable impact on energy performance yet, and perhaps understandably, building models still struggle to account for energy losses from poorly maintained building fabric and systems. In short, there remains a gap between getting a reasonable handle on the extent of disrepair of the building stock and producing robust figures of its impact on energy demand.

Producing more accurate figures would obviously come at a cost, and some experts argue that we should devote fewer resources to insulating homes and more to providing them with more heat from cheap and low-carbon sources, and whilst this argument has its merits, it neglects the immediate and cost-effective social benefits of improving the energy performance of the housing stock. However, as previously discussed, policy makers who advocate greater support for this 'fabric first' approach need to be particularly wary of the limitations of the models and the methods employed for data collection.

Only a small number of buildings are subjected to the rigorous (if not rigorous enough) on-site assessment necessary for a full SAP or SBEM assessment, and whilst the costs of surveying and the need for qualified assessors may seem prohibitive, and the models are somewhat flawed, there is currently no other sufficiently robust means of producing an accurate assessment of energy demand that meets the UK's reporting requirements. The Home Energy Questionnaires issued by the Energy Saving Trust and other organisations (and used for rdSAP assessments) are not without value, but that value is, arguably, limited to awareness raising and behaviour change. They may also have value in being able to collect data on maintenance issues, but this really needs verification by a qualified assessor to be sufficiently robust. An alternative method sometimes employed is to use surveyors to conduct drive-by surveys, but calculations using these data rely on assumptions based largely on building type and cannot capture maintenance issues that are not visible externally.

A corollary argument to this is that it is difficult to predict the impact of energy efficiency intervention programmes without an accurate baseline to begin with, and nor is it safe to assume that any installed measures will have similar impacts on different dwellings—for example due to technical problems or poor installation. However, whilst the barriers to assessing energy demand from existing domestic buildings largely stem from a lack of resources and investment, both currently and historically, there is no reason why new build could not be more accurately assessed, and whilst the rate of new domestic build and increasingly stringent building regulations may deter political support for greater assessment, there is a clear value to much greater assessment of both existing and new non-domestic build that also makes economic sense.

Finally, keeping heat in our buildings and knowing how well they retain it (or indeed ventilating it out when it's not wanted, and knowing the energy cost of doing that) is only one half of this problem. The other is getting it in there in the first place, and this brings us to what is arguably the biggest failing of both Scottish and UK energy policy—the lack of investment in infrastructure.

5.7 What Went Wrong?—not Investing in Infrastructure

Imagine this scenario. You are the latest civil servant to step off the civil service roundabout and into the world of energy and the built environment. You, in common with many of your colleagues, have little prior knowledge or experience of the field, and you may well have been trained in conventional social science-led approaches to policy-making that struggle to cope with highly complex problems (Castellani 2014). Your superiors are demanding speedy and low-cost solutions to the increasingly intractable problem of squaring the circle of reducing emissions from the built environment whilst also tackling fuel poverty and the legacy of poor maintenance that has driven it. And, on top of all that, they are being lobbied by a profit-driven construction industry that has still to fully emerge from the recession. What you really do not want to hear is that your priority should be to plough significant amounts of political capital into solutions that are not only costly but also long term, leaving the benefits open to being claimed by a future administration. Nor, probably, do you want to hear that the solutions have the potential to cause disruption to the public and markets whilst placing additional demands on the public and private sectors, and could help take the ownership of energy supplies further away from the hands of your lobbyists in the dominant energy companies. Yet, that is a reasonable summary of the position of anyone arguing for investment in energy infrastructure today, and the benefits that infrastructure could unlock. However, the long-term costs of not unlocking it means that this is a fight the Scottish Government cannot afford to lose.

But let us not pretend that this is anything new to either Holyrood or Westminster. If there is one political need on which even the most left-wing academics and right wing directors of industry invariably agree on it is that to leverage significant investment in energy infrastructure, governments must send long-term signals to the industries (recent examples include: Baker et al. 2012a; Ernst and Young 2012; RWE npower 2012). Without pipes and cables in the ground the capacity to adapt to future changes in energy supply and demand will be severely limited, and investment needs to be sufficient to ensure the most optimal pipes and cables can be chosen to future-proof the energy networks. For academics and policy makers, this means it would be possible to develop far more robust and accurate projections and longer-term policies to better manage the coming changes to the industry and building stock; and for the industry, it means being able to safeguard longer-term profits by installing more robust and flexible (but also costlier) infrastructure now, and thereby being able to plan for a more radically different and significantly lower carbon energy future.

Here again, there have been signs of hope—for example the pushing through of the Beauly-Denny power line in the face of stiff opposition—but they remain few and far between, and invariably for distributing electricity rather than heat. Furthermore, the Scottish Government has also frequently found itself in conflict with Westminster, for example over its setting of strike prices for nuclear and on and offshore wind, and the unswerving commitment to maintaining the status quo of privately owned centralised generation (HM Treasury 2013). Therefore, it is hardly surprising, and indeed welcome, that the Scottish Government's much-heralded white paper on independence emphasised the need for Scotland to use a yes vote and the country's significant renewable energy resources to forge a profitable and low-carbon energy future that is distinctly different than Westminster's plans for the rest of the UK (Scottish Government 2013c). However, whilst the white paper addressed energy (and fuel poverty) at length, it made little distinction between the provision of electricity and heat, and yet the development of more local and decentralised renewable heating networks is something the Scottish Government could already have been leading on regardless of the outcome of the referendum.

In this respect, the Scottish Government's Second Report on Proposals and Policies (RPP2—Scottish Government 2013d) was a critical missed opportunity to bring heat firmly onto the political agenda. Instead its proposals on heat were relegated to its separate Outline Heat Vision (Scottish Government 2013e), and its proposals on district heating further relegated to its District Heating Action Plan (Scottish Government 2013f), with the trail leading onwards to the Heat Policy Statement (Scottish Government 2015a). This statement sets out some important new measures, such as the revised Home Energy Efficiency Programmes, that are intended to better address energy efficiency, fuel poverty, and the effect of the urban–rural divide; however, these are still largely incremental changes. Furthermore, the statement was written before the Treasury's cutting of the Green Deal and restriction of ECO funding, and yet again a cornerstone measure, Scotland's Energy Efficiency Programme (SEEP), is deferred to 2017/18 whilst the

Scottish Government awaits the conclusions of the Smith Commission on greater devolution of powers. As an indicator of the level of optimism in the likely effectiveness of the SEEP, it has been memorably described by one well-known stakeholder, who shall remain anonymous, as '*a slow motion car crash with an indeterminate number of casualties*'.

As a brief aside, the term 'step change' is used frequently in energy policy and might just be the most abused term in Scottish politics, and the Outline Heat Vision provided an excellent example in relation to the deployment of district heating. The proposals promise that the recommendations of the Expert Panel on District Heating will '*help us work towards bringing about a step change in district heating in Scotland*'. This is a statement that merely committed the Government to convening a panel and responding to, but not necessarily taking up, its recommendations. As a statement of intent, it says nothing and means even less. This criticism could equally apply to the aspirational nature of the statements in the Scottish Government's Planning and Heat Statement (Scottish Government 2015b).

In reality, across all of these documents and beyond much of what accounts for a concrete strategy for heat relies on converting much of Scotland's heating demand to electricity. However, this demands an answer to the question of whether converting more and more buildings to electric heating is the best solution in the first place, and here there remains much research to be done. Aside from the far from insignificant costs, whether to consumers or to public or private purses, and the likelihood of Westminster-led programmes such as the Green Deal and ECO delivering sufficient numbers of conversions, there is also the question of how the performance of the building stock, particularly older and traditional buildings, will be able to physically adapt to the differing heating and cooling load patterns of electric and other alternative and supplementary heating systems (STBA 2012). Very little research has been done in this field (the STBA report finds just one study, conducted by Historic Scotland) yet as the Scottish building stock consists of high proportions of these buildings, so the potential costs of remedial work and/or re-conversion alone should justify adherence to the Precautionary Principle.

One exception here is solar thermal, which can be mounted on and integrated with existing buildings and heating systems with minimal disruption to the building fabric, yet its contribution remains negligible (9 GWh total output for 2010—Scottish Government 2012) and the multiple occupancy nature of many Scottish buildings remains one of several barriers to greater uptake (see Chap. 7). Yet again, another barrier is a lack of investment in infrastructure—in this case, for retrofitting shared buildings to enable occupants to capture this woefully under-used zero-carbon energy resource.

Alternatively, the Scottish Government could plan to deliver low- and zero-carbon forms of heating that minimise the risk to the existing building stock over the long term, whilst making more radical changes to drive new build towards passive and electric heating and cooling systems. This approach could yield cheap and low-to-zero carbon heating over the long term, as well as doing much to alleviate fuel poverty; however, by being heavily reliant on investment in new and upgraded infrastructure, it is unlikely to yield the rapid emissions reductions demanded by the targets it has set itself. This is not to say that those targets themselves are overly ambitious or unachievable, and nor that they are the root cause of the problem, but their political symbolism (and vulnerability) means they are often both abstractions and distractions from more constructive long-term policy-making.

5.8 The Perfect Storm Arises

Here lies the genesis of that perfect storm. The Scottish Government has grasped the huge potential of its renewable energy resources to provide vast quantities of relatively cheap electricity, but it also has its eyes on meeting the climate change targets and conducting a damage limitation exercise on the failed 2016 fuel poverty target. This has resulted in a political environment that is far from conducive to making the long-term commitments needed to give confidence to investors in non-electric renewable heating infrastructure; and with the economy and industry only beginning to emerge from recession this mounting impasse could not come at a worse time for the country.

This perfect storm is already on its way, thanks to years of short-term political thinking and too much belief in the power of ambitions and targets, but whether it hits Scotland with its full force or a more glancing blow could still depend on decisions made now. There might, just, be time to avert the worst, but the practicalities of building resilience through investing in infrastructure mean that it is close to running out, and the Scottish people cannot afford to wait for another referendum.

Scotland certainly has the potential to become '*the best wee country in the world*', but only if it can match its long-term ambitions with equally long-term evidence-based policy-making.

5.9 Conclusion: An Alternative Heat Vision

We urgently need a fundamental change in the way we see heat. If we could see it piling up in landfills, there would, rightly, be a public outcry over how much of this recyclable resource is being wasted. Unlike electricity, heat is fundamentally a waste product, and we produce and waste vast quantities of it.

When we see heat as a waste problem then the solutions become clear. The ultimate goal of waste management, after prevention, is to capture and recycle as much of a waste stream as possible, and to do so as efficiently as possible. If we apply this to heat, we get prevention through passive building design, energy efficiency and solar thermal, and recycling through district heating networks. Other technologies come into play but are essentially supplementary—in that they aid the efficiency of the system and improve the quality and quantity of the recycled product, but they are still inputs into that system. Governments have long since come to terms with the fact that regulation is essential for waste reduction, and the benefits of building regulations were understood as far back as Hammurabi (Harper 1904). So why do we not regulate waste heat?

At a time when well over a third of Scottish households are struggling to heat their homes, it seems scandalous to be building new power plants that are not equipped to capture waste heat and deliver it to those in need. Yet there is currently no law to prevent owners of generation plants doing just that, whether they are energy companies or public and private sector operators of smaller scale systems.

The proposals for a biomass plant at Leith Harbour, Edinburgh, which would have dumped its waste heat into the Firth of Forth, are just one of many examples of attempts to get away with practices which, if applied to solid waste, would be prevented by law. However, following a lengthy inquiry the proposal, which was memorably described by Marco Biagi MSP as '*the worst example of greenwash I've ever seen in Scotland*' (Tibbitt 2011), ultimately fell on the unsustainability of large-scale biomass, rather than the social unsustainability of siting a major source of waste heat next to a concentration of fuel poor households and not linking the two. Yet in Denmark, which has had a Heat Planning Law since

1979 (Danish Energy Authority 2005), this and other similar plants would never have made it past the initial planning application.

If the Scottish Government wants a mechanism for delivering a real step change in heat planning, and one which is not reliant on independence, then it need look no further than a Heat Planning Law. In its simplest form, a Heat Planning Law would require all developers of major sources of waste heat to install infrastructure to capture and deliver it to local homes and businesses from day one of operation, and conversely to compel all developers of new housing and non-domestic buildings to connect them to district heating systems (or to meet 100% of heating demand heat from other renewable sources). To smooth the transition, and to appease developers with projects already in the planning system, an introductory phase could be added in which developments without a nearby source or sink for heat would need fitting with infrastructure for subsequent connection (to borrow a phrase from the energy industry we could call them 'DHS ready'), but any more than that and the step change quickly becomes a gradual slope (Baker et al. 2012a).

However, let us not forget that in the short term, much of the fuel used for CHP and district heating systems will still come either from fossil fuels or biomass, which is at best unsustainable on a large scale. The next step would be converting those systems to alternative fuel sources, which could initially be biogas mixes (hence the need for much more flexible infrastructure as well as much more of it); combined with some sustainable biomass (sourced from sustainably managed Scottish forests); anaerobic digestion (which has huge potential in rural areas and urban margins); and possibly some contribution from energy from waste plants (Emmanuel and Baker 2012). So, until, and if, the hydrogen economy finally comes online CHP and DHS will remain largely low-carbon solutions, rather than renewable ones, but by being able to supply large quantities of cheap heat directly to those most in need, and by encouraging co-location of housing and employment, they offer the potential to deliver social and economic benefits that easily outweigh the cost of investment. But that potential is of course long term, so realising it requires the sort of political will and ambition that has so far been largely limited to targets and the easy win of large-scale renewable electricity.

Furthermore, district heating is far from the only solution to the Scotland's, sometimes conflicting, demands to meet its energy needs from renewables whilst also eliminating fuel poverty. As previously mentioned, Scotland's solar thermal resource represents another source of largely untapped cheap and renewable energy, particularly in urban centres with large areas of shared roof space. It might not seem like there is much of it on a dull winter's day in Aberdeen, but the cost-effectiveness of solar thermal panels as well as their ability to provide zero-carbon heat and the potential to integrate them with other heating systems (there's that infrastructure argument again) means they cannot be discounted as an energy solution. The same applies to solar photovoltaics, where efficiencies continue to rise as costs fall.

Yet another option that cannot be discounted are Air Source Heat Pumps (ASHPs), especially those with high-efficiency factors and integrated solar photovoltaics; however, the figures provided in the Second Report on Proposals and Policies and related documents suggest that the Scottish Government has overestimated their current potential. One likely source of this error may be the efficiency factors quoted by David MacKay in his ground-breaking book 'Sustainable Energy without the Hot Air', which quotes coefficients of performance of 4.9 and 6.6 for Japanese models (Mackay 2008). If so, then this serves as a classic example of what can happen when non-technically trained civil servants see a figure that fits their agenda and run with it. In reality, coefficients of performance for ASHPs in the UK have been found to be highly variable, but averaging around half of MacKay's figures, and despite their apparent political popularity the numbers installed to date have been far from inspiring (Baster 2011; EST 2010; WPZ 2011).

Other heat pump systems also have significant potential to meet the Scottish heating demand. At the small end of the scale, individual Ground Source Heat Pumps (GSHPs) are an option in rural and less densely populated areas, whilst larger community-scale variants could be used to mitigate the efficiency losses from using individual pumps in more densely populated urban areas. And on a much larger scale, the pioneering use of GSHPs to extract heat from the flooded mine-workings beneath the Shettleston housing development, designed by John Gilbert Architects, is now being developed further in Glasgow as part of a ground-breaking project that could meet 40% of the city's heating demand (BGS 2013; Church 2012). This and other projects are also beginning to tap the potential from other abandoned subterranean infrastructure such as disused rail tunnels and sewers (New Scientist 2013), and similar work is underway to tap deep geothermal (AECOM 2013).

However, most of these alternatives are already large scale, and scaling up to national levels, where the appropriate site-specific resources exist and have been mapped, will take time. In the short term, measures to manage heat demand need to be as much about reducing individual consumption and improving energy efficiency as finding new ways of generating it, and between generation and consumption lies distribution. This means infrastructure is the critical link in the chain that will link the energy-hungry yet wasteful built environments of today with the low carbon, efficient and sustainable built environments of tomorrow. So, investing in new and upgraded heating infrastructure, and regulating against wasting heat, could avert the worst of that perfect storm. But those steps demand long-term political thinking, and they need to be taken today—but are they a step too far for the Scottish Government?

Note

1. To its credit, the Scottish Government has also used the National Home Energy Rating Service's NHER model for domestic energy consumption, which was arguably a better predictor of energy demand, until its funding was withdrawn by Westminster.

References

AECOM, 2013. Study into the Potential for Deep Geothermal Energy in Scotland: Volume 1. Scottish Government publication (online) http://www.scotland.gov.uk/Resource/0043/00437977.pdf.

Affinity Sutton, 2013. FutureFit: Final Report Part 2. Affinity Sutton July 2013 (online) http://www.affinitysutton.com/media/364652/futurefit-quick-links-PDF-1.pdf

Baker, K.J., R. Emmanuel, and M. Phillipson. 2014. Review of the Energy Assistance Package. Communities and Analytical Services Division, Scottish Government.

Baker, K.J., R. Emmanuel, and M. Phillipson. 2012a. Support for RPP2—Housing Futures. Report for ClimateXChange Scotland. (online) http://www.scotland.gov.uk/Resource/0038/00389071.pdf.

Baker, K.J., R. Emmanuel, and M. Phillipson. 2012b. Support for RPP2—Abatement—Built Environment. Report for ClimateXChange Scotland.

Baker, K.J., and M. Rylatt. 2008. Improving the Prediction of UK Domestic Energy Demand using Annual Consumption Data. *Applied Energy* 85: 475–482.

Baster, M.E. 2011. Modelling the Performance of Air Source Heat Pumps. University of Strathclyde, Glasgow (online) http://www.esru.strath.ac.uk/Documents/MSc_2011/Baster.pdf.

Beckmann, K. and S. Roaf. 2013. Workshop Report: Climate Resilience for the Scottish Built Environment. ClimateXChange Scotland (online) http://www.climatexchange.org.uk/files/6113/7356/2210/CXC_Built_Env_EnablEnv_WorkshopRecommendationsReport.pdf.

BGS, 2013. Heat energy beneath Glasgow. British Geological Survey (online) http://www.bgs.ac.uk/research/energy/geothermal/heatEnergyGlasgow.html.

Bruhns, H., P. Steadman, and H. Herring. 2000. A Database for Modelling Energy Use in the Non-domestic Building Stock of England and Wales. *Applied Energy* 66: 277–297.

CAR, 2009. Modelling Greenhouse Gas Emissions from Scottish Housing: Final Report. Cambridge Architectural Research report for the Scottish Government (online) http://www.scotland.gov.uk/Publications/2009/10/08143041/0.

Castellani, B. 2014. Complexity and the failure of quantitative social science. Discover Society November 2014.

CFS, 2012. Consumer Focus Scotland's response to the Scottish Government Building Standards Division Consultations on: Section 63: Energy Performance of Non-Domestic Buildings; and Energy Performance of Building Directive—Recast. Consumer Focus Scotland (online) http://www.consumerfocus.org.uk/scotland/files/2012/01/Consumer-Focus-Scotland-response-to-SG-Building-Stds-Consultations.pdf.

Church, E.S. 2012. *A Proof of Concept Study Investigating the Application of a Geographical Information System in Determining Geothermal Potential in Abandoned Mine Workings in Glasgow*. Scotland: Glasgow Caledonian University.

Danish Energy Authority, 2005. Heat Supply in Denmark (online) http://193.88.185.141/Graphics/Publikationer/Forsyning_UK/Heat_supply_in_Denmark/pdf/varmeforsyning_uk.pdf.

DECC, 2013. Green Deal and Energy Company Obligation (ECO) statistics. Department for Energy and Climate Change, Westminster, UK (online) https://www.gov.uk/government/collections/green-deal-and-energy-company-obligation-eco-statistics.

EEP, 2013. Europe's Energy Portal (online) http://www.energy.eu/.

Emmanuel, R., and K.J. Baker. 2012. *Carbon Management in the Built Environment*. London: Routledge.

Ernst and Young, 2012. Powering the UK: Investing for the future of the Energy Sector and the UK (online) http://www.energy-uk.org.uk/publication/finish/5/298.html.

EST, 2010. Getting Warmer: A Field Trial of Heat Pumps. Energy Saving Trust publication (online) http://www.energysavingtrust.org.uk/content/

download/1801485/4898250/version/9/file/Getting_warmer_a_field_trial_of_heat_pumps_report.pdf.

Harper, R.F. 1904. *The Code of Hammurabi King of Babylon*, 2nd ed. Chicago/London: University of Chicago Press/Callaghan and Company.

Hills, J. 2012. Getting the measure of Fuel Poverty: DECC, Case Report 72 March 2012 (online) https://www.gov.uk/government/uploads/system/uploads/attachment_data/file/48297/4662-getting-measure-fuel-pov-final-hills-rpt.pdf.

HM Treasury, 2015. Our Energy Use (online) https://www.gov.uk/government/organisations/hm-treasury/about/our-energy-use.

HM Treasury, 2013. National Infrastructure Plan 2013 (online) https://www.gov.uk/government/uploads/system/uploads/attachment_data/file/263159/national_infrastructure_plan_2013.pdf.

IMechE, 2011. Scottish Energy 2020? Institution of Mechanical Engineers, London, UK (online) http://www.imeche.org/docs/default-source/2011-press-releases/IMechE_Scottish_Energy_Report.pdf?sfvrsn=0.

Jones Lang LaSalle, 2012. A Tale of Two Buildings: Are EPCs a true indicator of energy efficiency? Better Buildings Partnership (online) http://www.betterbuildingspartnership.co.uk/download/bbp-jll—a-tale-of-two-buildings-2012.pdf.

MacKay, D.J.C. 2008. Sustainable Energy without the Hot Air (online) http://www.withouthotair.com.

Maiden, T., K.J. Baker, and A. Faulk. 2016. Taking the Temperature: A Review of Energy Efficiency and Fuel Poverty Schemes in Scotland. Report for Citizens Advice Scotland by CAG Consultants, Glasgow Caledonian University, Andrew Faulk and the Energy Agency.

Mould, R., K.J. Baker, and R. Emmanuel. 2014. Behind the Definition of Fuel Poverty: Understanding Differences between the Fuel Spend of Rural and Urban Homes. Queens Political Review, II (2): 7–24.

New Scientist, 2013. How Heat from Trains and Sewers Can Warm Our Homes. 5th December 2013. Elsevier, London.

Restrick, S. 2013. Personal Communication. Scott Restrick, Technical Manager, Energy Action Scotland. Based on research conducted for: CFS, 2010. Turning up the heat: Benchmarking Fuel Poverty in Scotland. Consumer Focus Scotland (online) http://www.consumerfocus.org.uk/scotland/files/2010/10/Turning-up-the-Heat_final.pdf.

RWE npower. 2012. Energy and the Economy: The 2030 Outlook for Businesses. Report for RWE npower by the London School of Economics and the Grantham Research Institute for Climate Change and the Environment (online) http://www.npower.com/idc/groups/wcms_content/@wcms/@corp/@iac/documents/digitalassets/wcms_161561.pdf.

Sanders, C., and M. Phillipson. 2006. Review of Differences between Measured and Theoretical Energy Savings for Insulation Measures. Department for the Environment, Food and Rural Affairs (Defra), London, UK (online) http://www.decc.gov.uk/assets/decc/whatpercent20wepercent20do/supporting-percent20consumers/saving_energy/analysis/insulationmeasures-review.pdf.

Scottish Executive. 2002. The Scottish Fuel Poverty Statement (online) http://www.scotland.gov.uk/Resource/Doc/46951/0031675.pdf.

Scottish Government, 2009. Renewable Heat Action Plan for Scotland: A Plan for the Promotion of the Use of Heat from Renewable Sources (online) http://www.scotland.gov.uk/Publications/2009/11/04154534/0.

Scottish Government. 2010. Home Energy Schemes 2009–10: Energy Assistance Package and Home Insulation Scheme: End of Year Report.

Scottish Government. 2011a. 2020 Routemap for Renewable Energy in Scotland (online) http://www.scotland.gov.uk/Publications/2011/08/04110353/0.

Scottish Government. 2011b. Low Carbon Scotland: Meeting the Emissions Reduction Targets 2010–2022: The Report on Proposals and Policies (RPP1) (online) http://www.scotland.gov.uk/Publications/2011/03/21114235/0.

Scottish Government. 2012. Energy in Scotland: A Compendium of Scottish Energy Statistics and Information (online) http://www.scotland.gov.uk/Resource/0038/00389297.pdf.

Scottish Government. 2013a. Scottish House Condition Survey. Scottish Government (online) http://www.scotland.gov.uk/Topics/Statistics/SHCS.

Scottish Government. 2013b. Scottish Index of Multiple Deprivation. Scottish Government (online) http://www.scotland.gov.uk/Topics/Statistics/SIMD.

Scottish Government. 2013c. Scotland's Future: Your guide to an independent Scotland (online) http://www.scotland.gov.uk/Resource/0043/00439021.pdf.

Scottish Government. 2013d. Low Carbon Scotland: Meeting our Emissions Reduction Targets 2013–2027. The Second Report on Proposals and Policies (RPP2) (online) http://www.scotland.gov.uk/Publications/2013/06/6387/0.

Scottish Government. 2013e. Outline Heat Vision. Scottish Government, January 2013 (online) www.scotland.gov.uk/Resource/0041/00413386.pdf.

Scottish Government. 2013f. District Heating Action Plan: Response to the Expert Commission on District Heating. Scottish Government, May 2013 http://www.scotland.gov.uk/Resource/0042/00423849.pdf.

Scottish Government. 2014. Scottish House Condition Survey 2013—Key Findings (online) http://www.gov.scot/Resource/0046/00465627.pdf.

Scottish Government. 2015a. Heat Policy Statement (online) http://www.gov.scot/Resource/0047/00478997.pdf.

Scottish Government. 2015b. Planning and Heat (online) http://www.gov.scot/Resource/0048/00487284.pdf.

Snodin, H.M. 2008. Fuel Poverty in Great Britain, Germany, Denmark and Spain—Relation to Grid Charging and Renewable Energy. Report for Highlands and Islands Enterprise (online) http://www.hi-energy.org.uk/Downloads/General%20Documents/Report%20on%20Fuel%20Poverty%20in%20Relation%20to%20Grid%20Charging%20and%20Renewable%20Generation.pdf.

STBA. 2012. Responsible Retrofit of Traditional Buildings: A Report on Existing Guidance and Research with Recommendations. Sustainable Traditional Buildings Alliance (online) http://www.spab.org.uk/downloads/STBApercent20RESPONSIBLE-RETROFIT.pdf.

Tibbitt, A. 2011. Leith Biomass Update: Easter Greenwash Special. April 25th 2011 (online) http://greenerleith.org.uk/blog/category/energy-climate-change/page/9.

UKGBC. 2010. Zero-carbon Non-domestic Buildings. UK Green Building Council, March 2010.

WHO. 1990. Indoor Environment: Health Aspects of Air Quality, Thermal Environment, Light and Noise. World Health Organisation (online) http://whqlibdoc.who.int/hq/1990/WHO_EHE_RUD_90.2.pdf.

WPZ. 2011. Test Results of Air to Water Heat Pumps based on EN 14511. Wärmpumpen-Testzenstrum. Published by the Interstaatliche Hochschule für Technik Buchs, Switzerland (online) http://www.ntb.ch/fileadmin/Institute/IES/pdf/PruefResLW110620_Eng.pdf.

Author Biography

Dr. Keith Baker is a Researcher at the School of Engineering and the Built Environment, Glasgow Caledonian University and a co-founder of the Initiative for Carbon Accounting (ICARB). He specialises in energy in the built environment, with a particular focus on fuel poverty, energy efficiency and sustainable building materials.

CHAPTER 6

Scotland, Nuclear Energy Policy and Independence

Raphael J. Heffron and William J. Nuttall

6.1 Introduction

In 2010, UK government policy on energy provision reached a state of tension. One point of tension was that the parties comprising the new UK coalition government (the Conservatives and the Liberal Democrats) had expressed different manifesto commitments on energy. Second, in Scotland and Wales, the devolved governments each expressed the importance of developing large-scale renewable projects in their jurisdiction. Indeed, Scotland had moved further down its own path than Wales, and in 2008, the Scottish Parliament produced its own energy policy document for Scotland that included an expressed opposition to new nuclear power stations (Scottish Government 2008).

Despite the vote to remain within the UK in the 2014 Scottish Independence Referendum, a debate on energy policy in Scotland is still ongoing. Ever since the statement in 2008 in the earlier mentioned energy policy document where the Scottish Government said no to future nuclear new build in Scotland, it seems that debate on nuclear

R.J. Heffron (✉)
Queen Mary University of London, London, UK
e-mail: r.heffron@qmul.ac.uk

W.J. Nuttall
The Open University, Milton Keynes, UK
e-mail: william.nuttall@open.ac.uk

G. Wood and K. Baker (eds.), *A Critical Review of Scottish Renewable and Low Carbon Energy Policy*, Energy, Climate and the Environment,
DOI 10.1007/978-3-319-56898-0_6

energy as part of the Scottish energy mix has ceased. However, herein lies a significant problem. Nuclear energy provides *ca.* a third of Scottish electricity supply. Replacing this will be a significant problem and one that has not been addressed in any discussions yet.

Had Scotland chosen to go down the route to independence in the 2014 referendum, the process would still have taken some time. Alex Salmond, then leader of the SNP, was reported to have been keen to declare Scottish Independence in March 2016, just before the Scottish Parliamentary elections. This would have continued the generally nuclear-averse SNP policy of blocking new nuclear power stations, but the full range of issues shaping the future role of nuclear energy has not been explored. However, although the Scots narrowly voted to remain in the UK in 2014, the referendum process brought new energy to Scottish politics, and in the May 2015 general election, the SNP further increased their hold on Scottish politics by winning 56 of the 59 Scottish seats at Westminster. It remains to be seen how they will use this new influence, and as things stand today, the debate has centred on promoting the development of more wind farms and the ownership of the oil and gas resources in the North Sea. The continued focus on the extraction of oil and gas in the North Sea cannot contribute to Scotland developing itself into a low-carbon nation. Finally, the growing political divide between Scotland and the rest of the UK (as evidenced in the 2016 EU Referendum) and the likely continued dominance of the SNP means the prospect of a repeat referendum remains a distinct possibility, and so it would be an omission not to factor this into the discussions that follow.

In a strict constitutional sense, energy policy is not devolved from London to Edinburgh. The then UK Department of Energy and Climate Change formally has authority over policy in the whole of the UK, but the reality is that Scotland has much policy power in practice and the Scottish Government has an energy minister, Fergus Ewing. The strongest aspect of DECC policy power concerns electricity market arrangements on the island of Great Britain.[1] The British electricity market has included Scotland since April 2005 and the formation of the British Electricity Trading Transmission Arrangements which essentially brought Scotland into the market arrangements established in England and Wales. The (UK) Energy Act 2013 implements EMR and further adjusts market arrangements, especially as it relates to new build low-carbon generation. Despite the formal power of Westminster and Whitehall over Scottish energy policy, the reality, especially as relates to new infrastructure investment, is that the Scottish Government has power over all decision-making relating to environmental permission, and as such, it has a veto over all proposed investments. That reality ensures that while EMR will apply in Scotland, its powerful arrangements intended to make possible new nuclear build will have no direct impact in Scotland.

The current academic literature is somewhat divided on the issue of the effect of Scottish devolution on UK energy policy. The Scotland Act 1998 which created the new Scottish Parliament appears to have given Scotland extensive powers regarding the formulation of its own environmental policy (Little 2000). However, as Little (2000) determined, there are limiting factors such as the need to adhere to international and EU environmental law as well as the provisions for judicial review and political review by the UK Government. In this context, Keating (2010) argued that while Scottish devolution may be limited by intergovernmental relations with both the UK and the EU in many policy areas, there are nevertheless opportunities for policy innovation. Hence, while developing its own environmental policy may be a limited exercise, Scotland has in effect the capability to pursue its own energy policy. It remains to be seen how far Scotland may diverge in its approach to its energy policy from that of the residual UK. One of the distinctive differences with the rest of the UK has been the decision not to build new nuclear reactors. It is arguable therefore that energy policy has fragmented to a degree within the UK, and though this is a topic to consider in more depth in future, it can be stated that the Scottish move in 2008 to have its own energy policy has added to the uncertainty in the development of a long-term UK energy policy.

This chapter examines the role of nuclear energy in Scotland and the resulting concerns for Scotland as part of the wider independence referendum debate, a particularly relevant issue with the publication of draft legislation for another independence referendum just 2 years after the last one and the vote to leave the EU that was not mirrored in Scotland. The aim is not to provide an overview of the UK nuclear energy industry[2] nor engage in a discussion about a possible nuclear renaissance[3] but to focus directly on current Scottish energy policy and its relationship to nuclear energy. There are four central parts to this chapter: (1) the Scottish electricity mix is detailed; (2) a statement about nuclear energy by the Scottish energy minister is analysed; (3) nuclear energy as stated within the Scottish Independence White Paper is examined (Scottish Government 2013a); and (4) the issue of nuclear waste is assessed. The chapter does not purport to advise for or against Scottish Independence but aims to further the debate in an under-explored area of energy policy that will be of value whether Scotland secures independence or further devolution.

In this chapter, we also do not explore the interesting and contentious issues relating to nuclear weapons policy and an independent Scotland. The UK submarine-based nuclear deterrent is based at Her Majesty's Naval Base Clyde at Faslane with supporting infrastructures, such as the Royal Naval Armaments Depot Coulport nearby. In the event that Scotland were to become independent of the UK, a key infrastructure of the UK nuclear deterrent would be outside the UK. In October 2012, Scottish First Minister Alex Salmond spoke out against any connection between an independent Scotland and support for UK nuclear weapons. He is reported to have said: '*The UK government has two choices—they either relocate Trident to another part of the rest of the UK or alternatively they could use nuclear facilities in America or France*' (BBC News 2012a).

The civil nuclear power debate has parallels with the defence nuclear debate in discussions of Scottish Independence. There are also the possibilities of some linkages, such as that a slow erosion of Scottish civil nuclear expertise could have implications for the governance and even maintenance of the nuclear defence capabilities on Scottish territory. Notwithstanding such synergies, this chapter will focus entirely on civil energy policy issues.

6.2 The Scottish Electricity Mix

In 2008, the SNP leader and First Minister Alex Salmond asserted that Scotland had achieved electricity self-sufficiency even without the substantial nuclear power generation in the country (The Scotsman 2008). It would appear, however, that he was relying on annual averaging of just a number of years to justify his claim and periods of renewable power surplus (and export) were being allowed to offset periods of shortfall. At that time, on average, 20% of Scottish power was exported to England. In terms of the electricity market structure, even if there were a break-up of the currency union with the rest of the UK, it was anticipated that there would not be a significant impact in terms of investment or market withdrawal by energy companies.[4]

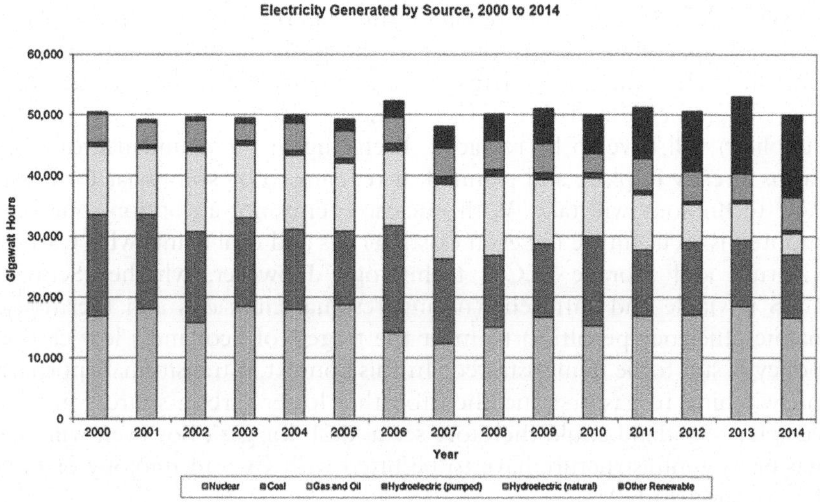

Fig. 6.1 Scottish electricity generation mix (*Source* Scottish Government 2016)

Data from 2002 to 2014 for the Scottish Electricity Mix are stated in Fig. 6.1. Notable is the reliance on fossil fuels that are 28% of the total electricity generation, and indeed, fossil fuels account for two-thirds of final energy consumption in Scotland (Scottish Government 2015). There is also a heavy reliance on nuclear energy, which provides 33.3%,[5] a high proportion considering that Scotland has a 'no to new nuclear energy' policy.

Scottish energy policy faces a challenge seen elsewhere in Europe (e.g. Germany). The stated goal of policy is to move towards a low-carbon economy. Alongside that ambition is a policy aversion to nuclear power, the policy of 'no to new nuclear energy'. In this constrained way, Scotland is aiming to change almost two-thirds or 61.3%[6] of its electricity generation. It is worth noting that in 2010, the proportion of fossil fuels and nuclear energy accounted for 79% of Scottish electricity generation, and the average over the period 2002–2014 is 80%. It is the Scottish Energy Policy of 2008 that dictated that all efforts would be made to embark on a path towards a carbon-free Scotland (Scottish Government 2008). More recently, this 'carbon-free Scotland' is also supported by the new Scotland 2020 Climate Group[7] and the *2020 Routemap for Renewable Energy in Scotland* (Scottish Government 2013b).

With the expected closure of the nuclear reactors[8] and the replacement of fossil fuel energy sources in the push for a low-carbon economy, it is clear that new energy infrastructure is needed, and therefore, the earlier mentioned 61.3% of Scotland's electricity generation (kWh supplied) will have to be replaced. Factoring in the abundance of wind farms already in place and planned, it remains to be seen what form this new technology will take. With nuclear energy not an option, one idea mooted is to fit future (or even existing) gas and coal plants with Carbon Capture and Storage (CCS) technology. However, whether Scottish CCS is viable and sufficient (noting residual emissions and thermodynamic efficiency penalties) to meet the targets of Scotland's low-carbon policy is still to be demonstrated. In this context, a transitional approach of switching to gas first and then to other lower-carbon sources may be adopted. Scotland could therefore see a 'dash for gas'; however, will new gas energy infrastructure have to be fitted with CCS technology or even be 'capture-ready'?

One of the main reasons for the Scottish 'no to new nuclear energy' is its failure to have a solution for its long-term waste. However, CCS technology has some similar waste issues to nuclear power (Reiner and Nuttall 2011). Furthermore, it is not presently commercially viable nor does it actually reduce the carbon emissions on a gas or coal power plant to zero. For example, gas produces just less than 400 kilograms per megawatt hour (kg/MWh) of CO_2, and through the use of CCS technology, these emissions will be reduced by *ca.* 85%; however, it will increase the cost of the construction up to 60% and reduce the efficiency of the power station (International Energy Agency 2013). Moves to fit CCS technology to Longannet power station in Fife, Scotland, stalled and were eventually scrapped in 2011 (BBC News 2011) with the power station itself closed on the 24 March 2016. Given these realities, and with an abundance of hydroelectric power plants (150 schemes currently) already in place, it remains for other renewable energy sources to help fill the looming supply gap in Scottish electricity supply. The strong presence of despatchable hydro-power in Scotland favours high levels of penetration by intermittent renewables, especially if there is a reduced obligation to address issues of grid instability in England.

The crux of the problem is the replacement issue of old energy infrastructure. There are challenges between the old and new energy infrastructure in Scotland. The legacy of environmental impact is not considered in sufficient detail when planning for new energy infrastructure. Simply because one technology choice is cheaper than another does not mean it actually is when viewed over its lifetime. Nuclear power stations have one of the longest lifespans (up to 60 years), while gas and coal plants have a shorter lifespan (up to 40 years) and wind turbines having in many cases just a 25-year life span.

There are also fuel and waste costs. In particular, the costs for waste (radioactive waste and carbon dioxide emissions) across the energy sector are an unresolved issue. The cost of waste for long-term storage for nuclear energy or CCS is difficult to quantify, while carbon emissions from coal-, gas- and oil-fired plants continue unabated and untaxed across the world. Wind projects suffer from reliability problems alongside interfering with wildlife and scenery. It is clear that whatever policy a government follows regarding energy infrastructure there will be both positives and negatives.

Due to the financial climate across the world, it seems more sensible to make decisions that can have a short- to medium-term impact. A low-carbon economy is highly desirable, and a government also needs to ensure the development of energy infrastructure that will not leave or create harmful environmental legacies for the next generation. However, at the same time it needs to ensure that a low-carbon Scotland does not come at too high a price for its current population. A transitional approach that may involve a Scottish 'dash for gas' might be one such strategy to achieve this. Overall, more modest goals that place climate change at the forefront of the triumvirate of policies—economic, environmental and energy—and decision-making will make for a more sustainable low-carbon economy in the long run.

6.3 Nuclear Energy in Scotland

Scotland has been home to nuclear power plants since their introduction in the UK, and there have always been issues with their location (Grimston et al. 2014). Currently, there are two nuclear power plants with two reactors each in operation in Scotland, see Table 6.1. Nuclear power plants face significant environmental challenges when decommissioned. A need exists, however, to maintain and develop new expertise in this area with the final four operational reactors in Scotland due to be closed by 2023, although Hunterston B was originally scheduled for closure in 2016.

Table 6.1 Nuclear energy plants in Scotland, UK

Name	*Capacity*	*Technology*	*Began operation*	*Scheduled for closure*	*Status*
Hunterston B	960 MW (current)	AGR	1976	2023	
Hunterston A	2 × 160 MW	Magnox	1964		Closed—1989
Torness	1185 MW (current)	AGR	1988	2023	
Chapelcross	4 × 49 MWe	Magnox	1959		Closed—2004
Dounreay	Research reactors	DMTR, DFR, PFR	1955		Closed—1994

Source Compiled by the authors and EDF (2013)

A nuclear power plant generally can take 4–6 years to build, but it can also take the same amount of time to actually begin the construction on the project. Reactor design approval, project finance and planning permission all take years to prepare and obtain. While the UK is in the process of making and ensuring significant gains in these areas, a nuclear new build project is still one with a long-term planning and development phase. Hence, a government decision on developing nuclear new build should be taken sooner rather than later. One important consideration in this context will be life extensions. Life extensions are extremely cost-efficient investments for the operating company. The economic attractiveness of a nuclear reactor life extension far exceeds that of new power station construction of any type, especially nuclear new build.

The energy company, EDF, now owns and operates the nuclear power plants in Scotland. Recently, they began to push for an extension of the operating lifespan of the nuclear reactors in Scotland. The Scottish Government has publicly stated that they will not object to these potential life extension requests (BBC News 2012b). In December 2012, Hunterston B received a life extension until 2023 (BBC News 2012c). In addition, the Advanced Gas-cooled Reactor (AGR) twin reactors at Torness (the other nuclear power plant location in Scotland) are among the most modern of the UK fleet of AGR plants. The presumption is that EDF will follow a strategy beyond 2023 of securing life extensions of seven-years duration (EDF 2012). EDF is currently developing this programme of securing life extensions for these ageing plants from beyond 2023; however, it may only be able to secure 1–2 more life extensions. Therefore, based on 7-year life extension, 2030 or 2037 may see both nuclear power plant closures in Scotland and a need for this

Table 6.2 Nuclear workforce by Scotland region 2013

Scottish region	*2013 percentage (%)*
Central Scotland	5.4
Glasgow	13.4
Highlands and Islands	27.5
Lothian	1.3
Mid Scotland and Fife	2.5
North East Scotland	0.1
South Scotland	24
West Scotland	25.7

Source Nuclear Industry Association (2013)

energy supply to be replaced. Hence, the importance of the government decision on developing nuclear new build is highlighted again.

The mutual importance between Scotland and the UK nuclear industry is revealed by the fact that, *ca.* 2013, the UK had roughly 63,000 people employed in the nuclear sector including the direct supply chain, of whom 8.3% are based in Scotland (Nuclear Industry Association 2013). While the Scottish Government is somewhat hostile to nuclear weapons, civil nuclear power and nuclear research, these various capabilities represent an important part of the Scottish economy, especially in some regions, such as the remote Caithness coast (Highlands and Islands, see Table 6.2) where the Dounreay laboratories are located. In Table 6.2, it is evident that the majority of the nuclear workforce are in relatively vulnerable employment regions, and post-independence, this may put some of these jobs under threat, especially when the 'no new nuclear energy' policy of the Scottish Government is factored in.

6.4 The Scottish Government and Nuclear Energy

Despite the current Scottish Government's anti-new nuclear build stance, nuclear energy is destined to play a role in its electricity sector until at least 2030 (see earlier comment on planned life extensions). It is not clear what will happen then and if at a later date there will be a review of the stance of the Scottish Government on nuclear energy. The nuclear energy option in the future in Scotland is an issue for further examination, and to date, it has received little attention. A brief look at the reasonably interested reader at available publications from the Scottish Government, NGOs and academic literature points towards an area that is under-analysed and researched. Furthermore, and equally significant, the contribution of nuclear energy to the overall energy policy debate seems misunderstood.

This chapter will next examine two recent (2013) communications by the Scottish Government in relation to nuclear energy. The first is a statement by the Scottish Energy Minister specifically on nuclear energy. The second analyses the references to nuclear energy in the Scottish Government independence document.

On the 23 October 2013, the Scottish Energy Minister Mr. Fergus Ewing (Scottish Nationalist Party and Minister for Energy, Enterprise and Tourism) stated the following in response to the UK Government's ambition in its announcement to support the development of a new nuclear plant:

> Today's announcement confirms that consumers across the UK will be paying for nuclear generation until after the middle of this century. The single nuclear station at Hinkley could be eligible for consumer funded payments totalling around £1 billion per year, depending on wholesale prices. These payments will apply for the length of the contract being awarded—which, at 35 years, dwarfs the 15 years being offered to renewable energy technologies.
>
> The cost of this single station alone is comparable to the £43 billion which the UK Government's budget is assigned for all energy technologies between 2013/14 and 2020/21 and risks squeezing out home grown developments for imported nuclear technology.
>
> This UK Government's misguided enthusiasm for nuclear comes at a time when other countries, such as Germany and EDF's home nation France, are either eliminating or scaling back their dependence on nuclear generation and when we should be putting the support to our renewables energy industry and the jobs it will support across the country.
>
> The guarantee of support and subsidy under this contract until after the middle of this century also sits in sharp contrast with the lack of a UK Government commitment to support our offshore renewables sector and its potential beyond 2020.
>
> The Scottish Government has an ambitious but achievable target to generate the equivalent of 100% of electricity from renewable sources by 2020,

alongside generation from thermal sources fitted with carbon capture and storage.

Nuclear energy cannot be relied on for our energy needs. The output from Scottish nuclear generation fell to historic lows in 2006 and 2007 due to unplanned outages. Although output has increased since then, nuclear generation has not yet recovered to its pre-2006 levels.

This underlines the susceptibility of nuclear to sudden interruptions, and supports the Scottish Government's drive towards a balanced energy portfolio, based on cleaner thermal generation and the advantages which our huge renewables potential offers to Scotland

Statement by Mr. Fergus Ewing, MSP, Minister for Energy, Enterprise and Tourism, 23 October 2013 (Scottish Government 2013c).

This statement has been chosen for analysis as it is one of the few communications verbal or written from the Scottish Government specifically on nuclear energy. The statement expresses a rather biased view against nuclear energy, at first in relation to the benefit of nuclear energy and second, in its discussion of the alternatives to nuclear energy. In addition, the statement demonstrates a disconnect between Scottish energy policy and Scottish economic policy of moving towards a low-carbon economy.

A major issue in the statement is the expectation that thermal generation power plants will have carbon capture and storage technology fitted to it. This is despite the withdrawal of funding to the Longannet project mentioned earlier. Back in 2011, £1 billion was promised to the carbon capture storage project by the UK government, but this proved insufficient to ensure project success (BBC News 2011). There were also parallel EU initiatives to support CCS, but unfortunately the UK and EU competition criteria were poorly aligned. There is also the question of the commercial viability of CCS. Such commercial viability is not expected until 2030 at the earliest (Lowe et al. 2010),[9] with the lack of commercial viability for CCS relating to the high cost of adding the CO_2 capture plant; transport infrastructure and storage operations; the impact of CO_2 capture on plant performance in output and efficiency; and the market structure of the electricity sector which does not cost CO_2 emissions properly. Even so, CCS technology does not decrease carbon emissions to zero but only by a percentage, and it uses energy from the fossil fuel plant itself to drive the processes, reducing overall power plant (fuel use) efficiency. A recent 2013 IEA report *Technology Roadmap: Carbon capture and storage* confirms this finding as stated earlier.

The statement shows a further misunderstanding of energy policy in that it cites the changes in both France and Germany regarding nuclear energy as support for criticising the decision of the UK government. This demonstrates a lack of understanding energy policy holistically. The energy policy of Germany is a contradiction. The movement against nuclear energy is largely determined by internal political opinion and not on more technical climate policy, or arguably even energy policy, considerations. Fundamentally, German energy policy and the 'Energiewende' may be characterised as a firm move away from nuclear power and in favour of renewables, smart grids and energy efficiency. Any low-carbon benefits are purported to emerge from this system transformation, but thus far they are not being seen. At this stage, one sees a collapse of German wholesale power prices and an erosion of the market value of power companies, while retail electricity prices are some of the highest in Europe because of a socialised component of the domestic consumer bill sitting atop the wholesale power component. Struggling power companies note the collapse of the EU Emissions Trading Scheme and the very low price on carbon emissions and hence direct investment to highly CO_2 polluting lignite- and hard coal- based generation capacity. New coal-fired power stations are being built in Germany, and German power sector emissions, although falling overall, exhibited some recent annual volatility (e.g. increases between 2009–2010 and 2012–2013) (Cleantechnica 2016). However, if the proposed new coal-fired power stations are built, then greenhouse gas emissions are likely to increase: Is this what a low-carbon Scotland is aiming for?

France is not necessarily scaling down its nuclear energy ambitions. Despite the statement by President Hollande that the intention of energy policy in France is to reduce nuclear energy to 50%, there are a variety of other factors at play. There was political motivation by the French President Francois Hollande to provide increased support for renewable energy and increase his political support from the Green Party (BBC News 2014). Hollande's government wants to reduce French reliance on nuclear energy while also aiming to increase significantly renewable energy sources, remaining averse to domestic 'fracking' and slowly removing its small reliance on fossil fuels. The French nuclear energy industry is also in transition, with a clear move to develop it globally rather than limit it to just France. The international nuclear energy industry is gathering pace, and EDF needs to use a significant amount of its resources abroad. There is also the advantage that licence renewals (life extensions) continue to be successful and France continues to be a net exporter of electricity. Nevertheless, France aims to keep half of its electricity supply from nuclear energy and its actions do not represent comparison with a 'no-nuclear policy'.

For Scotland, the answer to the question of how to achieve a low-carbon economy should focus on reducing and replacing its reliance on fossil fuels, which account for 28% of its electricity mix (see Fig. 6.1 earlier). This made all the more relevant with the closure of the 2.4 GW coal power station at Longannet in March 2016, signalling an end to coal-fired electricity generation in Scotland (Scottish Power 2016). The current focus on developing wind farms at a very large scale is not viable. Offshore wind farm development cannot be the sole solution as it is sometimes currently perceived, with recent cancellations of large offshore wind projects in the UK highlighting this issue.[10] The continued development of onshore wind farms is limited, and Heffron (2013a) refers to a *wind energy tipping point* where the public will demand that no more onshore wind projects are built. This would imply that the replacement of operating nuclear power plants by wind turbines is not a realistic option; i.e. *ca.* 10,000 wind turbines would be needed to replace the contribution of nuclear energy according to the data from the Energy Research Systems Unit (2013) at the University of Strathclyde.

An examination of public preferences in Scotland (Table 6.3) reveals that a significant proportion (32%) of the public would be in favour of nuclear energy being built in their area. All energy projects built near the public receive reasonable high public support, suggesting that people see economic benefits to any such energy project. More significantly is the 13% who would choose nuclear energy as the majority source for their electricity. This is quite high when compared to wind (at just 18%) and the total combination of fossil fuels (including shale gas) which is just 7% in total. This suggests a clear preference for low-carbon energy sources by the Scottish public, with 86% favouring the majority of their electricity supply coming from low-carbon energy sources.

In addition, there is the repeated assertion at the end of the statement that Scotland will benefit from cleaner thermal energy production. As stated earlier, it is highly suspect to base an energy policy on this because of the lack of CCS technology development. It emerges from this final

Table 6.3 Public preferences in Scotland in the UK in 2013

Energy Source	*Percentage in Support (%)*
A. Support for energy generation projects in their area?	
Hydro	80
Large scale wind projects	62
Solar	78
Nuclear energy	32
Bioenergy	59
Shale gas	24
Black oil	37
Gas (excluding shale gas)	42
Coal	34
B. From what source should the majority of electricity come from?	
Hydro	27
Large scale wind projects	18
Solar	15
Nuclear energy	13
Bioenergy	3
Shale gas	1
Black oil	0
Gas (excluding shale gas)	3
Coal	3

Source YouGov (2013)

part of the statement that Scotland plans to move towards a low-carbon economy with a continued ambition to use thermal/fossil fuel power plants; this is repeated in the White Paper on Scottish Independence.

The final paragraphs of the statement by Mr. Fergus Ewing, MSP, Minister for Energy, Enterprise and Tourism, reveal a policy bias against nuclear energy. Nuclear energy is apparently unreliable that there can be sudden interruptions and that Scottish energy policy should consist of a 'balanced energy portfolio, based on cleaner thermal generation and the advantages which our huge renewables potential offers to Scotland'. Nuclear energy's reported poor reliability stems from low levels of generation arising from minor outages at both plants in 2005 and 2006. The Minister's statement fails to mention the contribution of nuclear energy to base-load electricity provision and intermittency of renewable energy.

The fossil fuel industry plays a significant role in the Scottish economy. Indeed, there can be no doubt that they play an influential role in the political lobbying of various kinds in Scotland and down in Westminster.

A recent study in the USA demonstrated the scale of the fossil fuel lobby groups and calculated that they spend close to $900 million per year (Brulle 2013). It would be *naïve* to think that similar sums (in proportion to size of the industry and population) were not spent in the UK; indeed, recently Hutton (2014) expressed an opinion that lobbyists remain a problem in the UK.

6.5 Nuclear Energy in an Independent Scotland

An examination of material from the Scottish Government demonstrates similar contradictions (Scottish Government 2013a). The document, entitled *Scotland's Future: Your Guide to an Independent Scotland* (hereafter referred to as the *White Paper on Scottish Independence*), published in November 2013 makes for interesting reading regarding nuclear energy and its overall contribution to energy policy. Despite the referendum result, it seems unlikely that the SNP's thinking has changed much on this, as the aims stated therein remain consistent with Scottish Government policies.

It was a stated aim within the *White Paper on Scottish Independence* that an independent Scotland would aim to decarbonise its electricity supply (p. 18). It is not clear, however, despite the rhetoric in the document how Scotland would achieve this. The document highlights that one of three central aims is for Scotland to be a leader on climate change (p. 293) and that there will be an accelerated delivery of its commitments on reducing CO_2 emissions (p. 292). Yet, however, moving to discuss energy specifically, the *White Paper on Scottish Independence* repeats that the decarbonisation of the electricity sector is a priority and the continuation of its non-new nuclear stance (p. 299). The document states that renewable energy and its development are a safer and more cost-effective method of achieving this than investing in nuclear energy. The reader is not told, however, what will happen to the approximately 30% of electricity supply from fossil fuels. The implication is that an expansion of renewable energy and energy efficiency gains will render nuclear energy unnecessary.

The discussion on oil and gas (p. 301 onwards) makes for more revealing reading. The oil and gas reserves of Scotland will continue to be developed. Furthermore, an independent Scotland would support further exploration and as a result will continue with policies to support this. These policies include having tax-relief schemes and other incentives for oil and gas exploration (pp. 303–305). The oil and gas industry it highlights faces large upfront costs, and methods for lowering these and other costs such as decommissioning will be sought by the new Scottish administration.

The document specifies clear support for the export-oriented fossil fuel industry, while in contrast it criticises the UK government for supporting low-carbon nuclear energy. The global climate impacts of Scottish carbon making its way to the atmosphere are rather overlooked and assisted by the fact that international climate policy focuses on the geographical location of fossil fuel combustion not resource extraction. The *White Paper on Scottish Independence* holds nuclear energy as too expensive, in need of long-term contracts and being unsafe. These three latter issues will be examined in turn. The development of new nuclear energy is expensive mainly due to large upfront costs. It does not benefit from tax-relief schemes which apply to the oil and gas sector. A brief examination of the tax reliefs that oil and gas production companies receive is revealing. Tax reliefs are given for nearly every expense related to production, and exploration, for both planned, successful and unsuccessful projects. An overview is outlined in a document from the HM Revenue and Customs (2008) *titled A Guide to UK and UK Continental Shelf Life: Oil and Gas Taxation 2008*. In addition, UK taxpayers will also pay for decommissioning in the oil and gas sector which will now receive tax incentives on decommissioning costs, estimated at £30 billion over the next 15 years; this tax relief was granted in 2012 (Burges Salmon 2013). The offering of tax relief for development and decommission represents similar long-term contracts to those being given to nuclear energy in the UK. Part of the purpose of tax legislation for fossil fuels is stated as to '*allow a project to rapidly recover its costs*' (HM Revenue and Customs 2013). Why are low-carbon energy sources not treated the same way and allowed to recover costs of a project rapidly? Fundamentally, energy policy and oil and gas extraction policies occupy different worlds, and there is little joined-up decision-making. An independent Scotland is likely to do nothing to alter that reality as there are political benefits in preserving it.

Finally, in terms of safety, it is not specified in the *White Paper on Scottish Independence* why nuclear energy is unsafe. Safety is a key concern across the energy sector, and low safety standards in the energy sector are a subsidy as it reduces the cost of providing safe operations. If safety is seen in terms of fatalities, a recent study has shown that between 1971 and 2009 the use of nuclear power in comparison with fossil fuels has saved 1.8 million lives (Pushker et al. 2013). Fossil fuels in general have high fatality rates, for example, in the USA and China where there are *ca.* 30 and 2000 fatalities *per annum*, respectively.[11] The UK has suffered its own major accidents with Piper Alfa causing 167 deaths in 1985. The question arises, has the safety culture improved or do lower safety standards contribute to the lower cost of fossil fuels?

Douglas (2002: xix) makes an interesting connection in this regard and links big industry and government together, alongside the influence of political affiliation—in many ways the fossil fuel issue and not nuclear energy in Scotland mirrors this quote:

> Dangers are manifold and omnipresent. Action would be paralysed if individuals attended to them all; anxiety has to be selective. We drew on the idea that risk is like a taboo. Arguments about risk are highly charged, morally and politically. Naming a risk amounts to an accusation. The selection of which dangers are terrifying and which can be ignored depends on what kind of behavior the risk-accusers want to stop. Not risky sports, not sunbathing nor crossing the road; it was to do with nuclear or chemical hazards – in short, big industry and government. Subsequent survey research showed that political affiliation was the best indicator of the distribution of attitudes to risk.

6.6 Nuclear Waste in an Independent Scotland

This issue of nuclear waste is a problem in many countries worldwide. Perhaps the most forward-thinking countries on this issue are Sweden and Finland. Both these countries have had long-term nuclear waste storage plans and both are in the process of receiving final approval to begin construction (Heffron 2013b).

Scotland currently has two nuclear power stations with two reactors on each site. An independent Scotland might be expected to have to deal with the problem of Scottish nuclear waste and perhaps also the nuclear wastes arising from its plants that are currently being stored at Sellafield in England. The problem arises because under EU law it is the responsibility of member states to manage spent fuel and radioactive waste. However, it is possible that two or more member states can agree to use a common disposal facility under strict conditions (Europa 2016). So, Scotland could avail of the latter upon becoming an EU member state, assuming the residual UK were to agree.

In addition, EU law now obliges member states to develop national programmes for nuclear waste disposal and to notify them to the European Commission by 2015 at the latest. These national programmes must include a timetable for the construction of disposal facilities, as well as a description of the activities needed to implement disposal solutions, costs assessments and a description of the financing schemes (Europa 2016: in particular, see Chap. 2 Obligations, Article 5 National Framework (a)–(h)).

There is a conflict here in that Scotland has rejected the idea of a deep geological disposal facility (and also new nuclear build) (Scottish Government 2008: 7; 2011). Heffron et al. (2013)[12] comment that:

> Jamie Reed MP (Labour) voiced concerns in parliament in 2012 as to whether the government will commit itself to an analysis of the volumes of Scottish higher activity radioactive waste which is stored in England, the costs to remove them, where they will be located in Scotland, and who will be responsible for them in the long term. A Scottish government statement in reply suggested that proposals for nuclear decommissioning in an independent Scotland would be covered in a white paper to be published in November 2013.

However, the storage of nuclear waste and nuclear decommissioning has received little attention in the *White Paper on Scottish Independence*. It is addressed very briefly (pp. 520–521), and no solutions are given, just an expression that nuclear waste will be managed safely and effectively. The promised answers have not yet materialised.

6.7 Conclusion: The Nuclear Future in Scotland

Whether one is for, against or indifferent to new nuclear energy development, the topic highlights a major gap in Scotland's energy and environmental policy goals. The energy policy debate from the Scottish Government perspective has been reduced to a low-carbon energy development debate between nuclear energy and renewable energy. The challenge should be how to reduce the significant contribution of the electricity supply sector that comes from fossil fuels, noting Scotland's continuing enthusiasm for fossil fuel extraction and processing.

This continuation of the use of fossil fuels will continue the emission of carbon dioxide. This prompts the question, were an independent

Scotland to join the EU, would their accession agreement require them to close their fossil fuel power plants? For many new EU member states since 2005, the closure of fossil fuel plants has been a key part of meeting their accession agreements.[13] In addition, the European Industrial Emissions Directive (Directive 2010/75/EU) places limits on the emissions of fossil fuel power plants (in particular, sulphur dioxide (SO_2) and nitrogen oxides (NO_x).

For Scotland to aspire to being a low-carbon economy, to decarbonising its electricity market and to being a leader within the climate change community, it needs to tackle the issue of how to stop the continuation of burning fossil fuels. This is not something the Scottish Government has yet achieved. Until this happens, these policy aspirations of the Scottish Government must be seen as consisting of simultaneous enthusiasm for a renewables-led energy policy and an oil and gas extraction industrial policy. Between these two odd components sits nuclear energy, an important Scottish industry destined for neglect and decline almost irrespective of Scottish Independence.

Notes

1. Northern Ireland is part of a single electricity market with the Republic of Ireland.
2. Please see the following for analysis of history of the UK nuclear energy sector: Taylor (2007).
3. Please see the following for more detail on the nuclear renaissance: Nuttall (2005).
4. This is because expectation would be for minimal change due to EU policy being for further integration of electricity markets. In this context, Scotland could sell electricity to other markets just as the rest of the UK could buy electricity from other markets. Therefore, keeping a common electricity market would be reasonable for both sides despite no currency union. For more on this see UK Parliament Energy and Climate Change Committee, June 2012. This includes data on current investments by a number of energy companies, and their investment indicates that the prospect of the break-up of the UK currency union post-Scottish Independence is not such an issue for energy investment (Energy and Climate Change Committee 2012).
5. Based on 2014 data.
6. This comprises of the 28% of the total electricity generation by fossil fuels and the 33.3% from nuclear energy.

7. For more on the Scotland 2020 Climate Group, see the 2020 Climate Group (2014).
8. The closure of the two nuclear power stations in Scotland is not that straight forward, with long-term SNP support for extending the scheduled closure dates: due to be decommissioned in 2023, Electricitié de France (EDF) has extended operations until 2030. Life extensions have also been decided for nuclear stations in the rest of the UK (World Nuclear News 2016).
9. It is unclear when CCS will be commercially viable due in part to the fact that it is not yet known which is the best technology for CCS.
10. See recent media reports on two wind farm projects and one nuclear energy project: 'Plans for £5.4bn Argyll Array offshore wind farm near Tiree dropped' (BBC News 2013; Financial Times 2013) and the recent legal challenge by RSPB leading to the Court of Session quashing previously granted planning consent for four offshore wind farms with a combined capacity of 2.3 GW in Scotland (Scotsman 2016).
11. This is an average taken from across multiple sources from the US Labour Department to OECD statistics and is a conservative estimate. For a full list of sources please contact the author.
12. See also the following newspapers: News & Star (2012) and the Scottish Express (2012).
13. New entrants to the EU had to reduce state aid to the fossil fuel sector and also reduce emissions such as those from fossil fuel plants, for example: SO_2 NO_x VOC and NH_3. For more detail see the EU Accession Articles on Energy and the Environment respectively (European Parliament 2016).

References

BBC News. 2011. *Longannet Carbon Capture Scheme Scrapped* (online). http://www.bbc.co.uk/news/uk-scotland-north-east-orkney-shetland-15371258. Accessed 8 Jan 2014.

BBC News. 2012a. *Nuclear Weapons 'Outlawed' in an Independent Scotland, Says Salmond* (online). http://www.bbc.co.uk/news/uk-scotland-scotland-politics-20020839. Accessed 18 Apr 2014.

BBC News. 2012b. *Scotland's Nuclear Stations Could Stay Open* (online). http://www.bbc.co.uk/news/uk-scotland-scotland-business-17266084. Accessed 8 Jan 2014.

BBC News. 2012c. *Hunterston B Nuclear Power Plant will Run Until 2023* (online). http://www.bbc.co.uk/news/uk-scotland-glasgow-west-20590915. Accessed 8 Jan 2014.

BBC News. 2013. *Plans for £5.4bn Argyll Array Offshore Wind Farm Near Tiree dropped* (online). http://www.bbc.co.uk/news/uk-scotland-scotland-business-25364699. Accessed 27 Feb 2014.

BBC News. 2014. *France Struggles to Cut Down on Nuclear Power* (online). http://www.bbc.co.uk/news/magazine-25674581. Accessed 29 Mar 2014.

Brulle, R.J. 2013. *Institutionalizing Delay: Foundation Funding and the Creation of US Climate Change Counter-Movement Organisations.* Climatic Change, (Advance Access) doi:10.1007/s10584-013-1018-7.

Burges Salmon. 2013. *Guaranteeing Tax Relief for Decommissioning.* Oil and Gas Connect, Issue 9, August 2013 (online). http://www.burges-salmon.com/Sectors/energy_and_utilities/Oil%20and%20Gas/Publications/Guaranteeing_tax_relief_for_decommissioning_what_will_it_mean.pdf. Accessed 30 Nov 2014.

CleanTechnica. 2016. *German Greenhouse Gas Emissions Fell 4.6%*. In 2014, New Data Shows—February 4th, 2016 (online). https://cleantechnica.com/2016/02/04/german-greenhouse-gas-emissions-fell-4-6-in-2014-new-data-shows/. Accessed 15 June 2016.

Climate Group. 2014. *Scotland's Climate Group* (online). http://www.2020climategroup.org.uk/.

Douglas, M. 2002. *Purity and Danger: An Analysis of Concepts of Pollution and Taboo (Routledge*, Classics ed. London, UK: Routledge.

EDF. 2012. *Regulatory Story: Life Extensions* (online). http://www.edfenergy.com/about-us/shareholder-information/documents/Life_Extensions,_HPB_and_HNB_-_04.12.12.pdf. Accessed 18 Apr 2014.

EDF. 2013. *EDF Energy's Nuclear Power Stations* (online). http://www.edfenergy.com/energyfuture/edf-energys-approach-why-we-choose-new-nuclear/current-nuclear-sites. Accessed 8 Jan 2014.

Energy and Climate Change Committee. June 2012. *The Impact of Potential Scottish Independence on Energy and Climate Change* (online). http://www.publications.parliament.uk/pa/cm201213/cmselect/cmenergy/writev/scottish/m01.htm. Accessed 15 Apr 2014.

Energy Systems Research Unit. 2013. *University of Strathclyde, Scotland UK* (online). http://www.esru.strath.ac.uk/EandE/Web_sites/01-02/RE_info/interesting.htm. Accessed 8 Jan 2014.

Europa. 2016. *Council Directive 2011/70/Euratom OJ L199/48. Establishing a Community Framework for the Responsible and Safe Management of Spent Fuel and Radioactive Waste.*

European Parliament. 2016. *Treaty of Accession.*

Financial Times. 2013. *Wind Energy Faces Adverse Conditions as RWE Axes Project* (online). http://www.ft.com/cms/s/0/08fc5494-5686-11e3-ab12-00144feabdc0.html?siteedition=uk#axzz2lmSbh6tF. Accessed 15 May 2014.

Grimston, M., W.J. Nuttall, and G. Vaughan. 2014. The Siting of UK Nuclear Reactors. *Journal of Radiological Protection* 34 (2): R1–R24.

Heffron, R.J. 2013a. Accommodating Energy Law within Environmental Law: An Irish Exploration. *Irish Planning and Environmental Law* 20 (2): 56–64.

Heffron, R.J. 2013b. Nuclear Energy, Year in Review. In *Yearbook of International Environmental Law*, ed. O.D. Fauchald, D. Hunter, and X. Wang, vol. 23, 269–273. Oxford University Press: Oxford, UK.

Heffron, R.J., M. Allen, and D. McCauley. 2013. The Forgotten Law and Policy Issue: Nuclear Waste Management in Scotland. *Edinburgh Law Review* 17 (3): 325–332.

HM Revenue and Customs. 2008. *A Guide to UK and UK Continental Shelf Life: Oil and Gas Taxation 2008* (online). http://www.hmrc.gov.uk/international/ns-fiscal3.htm. Accessed 30 Nov 2014.

HM Revenue and Customs. 2013. *Guide to the North Sea Fiscal Regime* (online). http://www.hmrc.gov.uk/oilandgas/guide/prt.htm. Accessed 30 Nov 2014.

Hutton, W. 2014. Power Lobbyists and Fawning Ministers are Corroding Society. *The Observer*, 11 January 2014. http://www.theguardian.com/commentisfree/2014/jan/11/corrosive-influence-big-business-lobbyists. Accessed 12 Jan 2014.

International Energy Agency. 2013. *Technology Roadmap: Carbon Capture and Storage*. Paris, France: OECD/IEA.

Keating, M. 2010. *The Government of Scotland: Public Policy Making After Devolution*, 2nd ed. Edinburgh, UK: EUP.

Kharecha, Pushker A., and James E. Hansen. 2013. Prevented Mortality and Greenhouse Gas Emissions from Historical and Projected Nuclear Power. *Environmental Science Technology* 47 (9): 4489–4895.

Little, G. 2000. Scottish Devolution and Environmental Law. *Journal of Environmental Law* 12 (2): 55–174.

Lowe, A., B. Beasley, and T. Berly. 2010. Carbon Capture and Storage (CCS) in Australia. In *Carbon Capture: Sequestration and Storage*, ed. E.E. Hester, and E.M. Harrison. Cambridge, UK: The Royal Society of Chemistry.

News & Star. 2012. *Cumbrian MP Says Scotland Must Take Waste Back Under Independence*, 18 Oct 2012 (online). http://www.newsandstar.co.uk/news/business/cumbrian-mp-says-scotland-must-take-waste-back-under-independence-1.1005981?referrerPath=home. Accessed 8 Jan 2014.

Nuclear Industry Association. 2013. *Nuclear Industry Association—Industry Jobs Map* (online). http://issuu.com/nuclear_industry_association/docs/jobsmap_scot13. Accessed 29 Mar 2014.

Nuttall, W.J. 2005. *Nuclear Renaissance: Technologies and Policies for the Future of Nuclear Power*. Oxon, UK: Taylor and Francis.

Reiner, D.M. and W.J. Nuttall. 2011. Geological Disposal of Carbon Dioxide and Radioactive Waste: Similarities and Differences. In *Geological Disposal of Carbon Dioxide and Radioactive Waste: A Comparative Assessment*, ed. F.L. Toth. Springer: New York.

Scottish Express. 2012. *Nuclear Waste Bill Threat to Scotland*, 22 January 2012 (online). http://www.express.co.uk/news/uk/297168/Nuclear-waste-bill-threat-to-Scotland. Accessed 8 Jan 2014.

Scottish Government. 2008. *Scottish Energy Policy: An Overview*.

Scottish Government. 2011. *Scotland's Higher Activity Radioactive Waste Policy*. DPPAS11098 (01/11), 4–5.

Scottish Government. 2013a. *Scotland's Future: Your Guide to an Independent Scotland*. The Scottish Government: St. Andrew's House, Edinburgh, UK.

Scottish Government. 2013b. *2020 Routemap for Renewable Energy in Scotland—Update*, 19 Dec 2013.

Scottish Government. 2013c. *Scotland's Response on Nuclear Energy* (online). http://news.scotland.gov.uk/News/Scotland-s-response-on-nuclear-power-553.aspx. Accessed 8 Jan 2014; and for a profile specific to Mr. Fergus Ewing see: http://www.scotland.gov.uk/About/People/14944/Scottish-Cabinet/fergusewing.

Scottish Government. 2015. *Energy in Scotland 2015* (online). http://www.gov.scot/Resource/0046/00469235.pdf. Accessed 28 Oct 2016.

Scottish Government. 2016. *High Level Summary of Statistics Trends—Energy* (online). http://www.gov.scot/Topics/Statistics/Browse/Business/TrendData. Accessed 15 Oct 2016.

Scottish Power. 2016. *Longannet Closure Marks the End of Coal-Fired Electricity Generation in Scotland* (online). http://www.scottishpower.com/news/pages/longannet_closure_marks_the_end_of_coal_fired_electricity_generation_in_scotland.aspx. Accessed 15 Oct 2016.

Taylor, S. 2007. *Privatization and Financial Collapse in the Nuclear Industry—The Origins and Cause of the British Energy Crisis of 2002*. London, UK: Routledge.

The Scotsman. 2008. *Salmond: Nuclear Redundant in Self-Sufficient Scotland*, 10 January 2008 (online). http://www.scotsman.com/news/salmond-nuclear-redundant-in-self-sufficient-scotland-1-1073316. 27 Feb 2014.

The Scotsman. 2016. *Offshore Wind Farms Dead in the Water After RSPB Wins Legal Battle*—Tuesday 19 July 2016 (online). http://www.scotsman.com/news/environment/offshore-wind-farms-dead-in-the-water-after-rspb-wins-legal-battle-1-4181368. Accessed 28 Oct 2016.

World Nuclear News. 2016. http://www.world-nuclear-news.org/. Accessed Dec 2016.

YouGov. 2013. *Scottish Renewable Survey Results*, February 25th–26th Results (online). http://d25d2506sfb94s.cloudfront.net/cumulus_uploads/document/vj66wakgzm/YG-Scottish-Renewables-Archive-results-260213-renewable-energy.pdf. 29 Mar 2014.

Authors' Biography

Raphael Heffron is Jean Monnet Professor in Energy and Natural Resources Law & Policy at Queen Mary University of London, Co-Chair of the UK Energy Law and Policy Association, Visiting Professor in Energy Law at the International Hellenic University (Greece); and an Associate Researcher at the Energy Policy Research Group at the University of Cambridge where he read his Ph.D. A trained Barrister-at-Law, Raphael's research interests are in energy, environmental and planning law and policy with a focus on energy infrastructure development, electricity markets, energy subsidies, energy liability, the EIA process, energy justice and Arctic energy law. He has over 80 articles, books and publications of different types.

William Nuttall is Professor of Energy at The Open University, UK. His career has taken him from experimental physics to technology policy with an emphasis on nuclear energy policy. He is a Fellow of Hughes Hall, University of Cambridge. He has extensive public policy experience assisting the UK Government and the European Union.

PART II

The Challenges Ahead

CHAPTER 7

Reducing Demand: Energy Efficiency and Behaviour Change

Keith Baker

7.1 Introduction

When Plato wrote that *'Human behaviour flows from three main sources: desire, emotion, and knowledge'*, it is notable that he put knowledge last. Over 2000 years later, the actor Johnny Depp summed up how little things have changed with another quotable quote, *'People say I make strange choices, but they're not strange for me'*. And therein lies the most intractable problem for those of us who chose to specialise in the dark arts of energy efficiency, behaviour change and demand-side management.

From a technical point of view, the Scottish building stock poses a wide range of problems for improving energy efficiency, most obviously the high proportion of traditional tenements and other multiple occupancy buildings, but also the large numbers of traditional buildings in rural areas that require more individual attention (Scottish Government 2015). More fundamentally, the stock contains higher proportions of buildings in disrepair and those classified as 'hard to treat' than the rest of the UK, a problem which has been exacerbated by the loss of the skills and industries needed to maintain them (Roaf et al. 2008).

K. Baker (✉)
School of Engineering and the Built Environment,
Glasgow Caledonian University, Glasgow, UK
e-mail: keith.baker@gcu.ac.uk

G. Wood and K. Baker (eds.), *A Critical Review of Scottish Renewable and Low Carbon Energy Policy*, Energy, Climate and the Environment,
DOI 10.1007/978-3-319-56898-0_7

The current state of the Scottish building stock is not an issue that can be pinned on any one administration. It is a legacy of political footballs kicked about by successive administrations being unwilling or unable to develop an effective, coordinated and consistent response that spans the actions of numerous administrations and two governments. Its most recent incarnations, the abject failures of the Green Deal and the Energy Companies Obligation (ECO), may have been Westminster-led, but if energy policy was completely devolved, the Scottish Government would open itself up to more probing questions about its longer term intentions for reducing emissions from the built environment.

First of all, should we be talking about energy efficiency at all? If our real aim is to reduce emissions then energy efficiency means little if total demand is unaffected, and especially if the bulk of that demand must still be met from non-renewable resources. However, talking about improving energy efficiency is politically (and perhaps publicly) more palatable because it frames the problem as 'doing more with less', whereas 'demand reduction' evokes deeper green thinking and leaves the door open to those who would dare to challenge the Scottish Government's mantra of *'sustainable economic growth'*. Yet many experts argue that greater prosperity and a healthier society could be better achieved by sacrificing the sacred cow that is 'growth is good' (for example, Jackson 2009), and, as we will see later, there are other sacred cows that may need to be sacrificed on the altar of energy efficiency if the Scottish Government is to achieve its emissions-reduction targets.

In order to be confident of achieving Scotland's emissions targets, the longer term goal needs to be achieving more radical attitudinal and lifestyle change both at home and at work. Policies will need to impact far beyond the information-rich and environmentally aware, who are already using smart meters, installing solar panels, meeting by teleconference and cycling to work. They will need to reach the vulnerable and hard to reach, some of whom must be supported in increasing their energy use in order to maintain warm and healthy homes, and they must address those who are actively opposed to reducing their energy consumption. They also need to avoid disproportionately impacting on poorer households, indeed they must be part of the solution to poverty, whilst also having the teeth to penalise the obstinate and energy profligate. And that is perhaps the greatest challenge of all—designing policies for demand reduction that are sensitive to the fact that humans are innately human.

7.2 Technical Solutions to Improving Energy Efficiency

The UK's road to energy efficiency is paved with mediocrity. Where progress has been made, it has invariably been the result of tightening the Building Regulations (known in Scotland as the Building Standards), but the slow turnover of the building stock means that progress has been incremental and every opportunity not taken has left more to be done by retrofitting the 80% of buildings that will still be in use in 2050.

Since the privatisation of the energy industry government policies has revolved around trying to recoup some of the profits made by suppliers by investing them in schemes such as the Energy Efficiency Commitment (EEC) and the Carbon Emissions Reduction Target (CERT), supported by the work of local authorities and a limited range of grants. Whilst there have been some success stories, none of these have been notable at an international level, and more recently the Green Deal and ECO have set a world-class example of how to get things wrong. As regards the former, depending on whether you count the 'soft' launch in late October 2012 or the 'official' launch at the start of January 2013 DECC's own figures show that it clocked up either five or twelve installations in its first 6 months, and either 757 or 1312 in its first year (DECC 2015). Despite all the hype and a staggering amount of spin, the Coalition Government singularly failed to deliver on what should have been an incredibly easy task—selling subsidised insulation to the middle classes—before quietly pulling both into the remit of the Treasury in the summer of 2015 and then scrapping the Green Deal entirely.

The Energy Companies Obligation fared better, delivering just under 380,000 measures by the end of October 2013, but of these almost one-third were boiler replacements or repairs, another third were cavity wall insulations, and another third were loft insulations (amounting to 91% of the total). The remainder consisted of other forms of heating and insulation but, as of 2015, not one single installation of a microgeneration system (DECC 2015). And whilst ECO delivered significantly more measures than the Green Deal, this cannot be taken as political capital as ensuring these households have working boilers and homes that do not leak most of the heat they produce is at least as much a basic welfare issue as an environmental one. The figures were so shockingly bad that the only question that should be asked is whether this was down to simple incompetence or deliberate incompetence in an attempt to undermine the wider agenda. However, for the sake of argument let us have a quick run through some of the reasons why many experts predicted this spectacular failure from the moment the initial plans were announced.

To begin with, expert advice was ignored from the start. As just one example of how unified building and construction experts were in their opposition to the initial proposals, following a presentation by a representative of DECC at conference attended by over 200 professionals and academics in February 2012, their suitability for Scotland was put to the vote (CICStart 2013). Not one delegate indicated their support, and fewer than ten indicated that they thought the consultation exercise would lead to sufficient revisions. Yet despite the wealth of advice and warnings, the Scottish Government continued to assume that the Green Deal and ECO would deliver the projected benefits and incorporated these assumptions into key documents such as the second Report on Proposals and Policies (see Baker et al. 2012 and sources therein). Castles were built on sand, and with the political hiatus surrounding the referendum, the environmental, social and economic costs of inaction continued to mount up.

Then following the vote to remain in the union, the cancellation of the Green Deal and restriction of ECO funding, the relative impotence of the Smith Commission, the foundations of much of the Scottish Government's energy efficiency policies are now crumbling away from underneath it. So, whilst the governing SNP can take no small amount of credit for squeezing more funding out of ECO than the Treasury may have wanted to release, and using it better (Maiden et al. 2016), in doing so they gloss over the fact that the overall pot continues to shrink.

As an example of the naivety of that thinking, let us turn briefly to that cornerstone of the Green Deal, the 'Golden Rule' that all measures installed would pay for themselves in 40 years. If the UK's then Department of Energy and Climate Change had really believed in the figures it was pedalling, then it could've underwritten the savings itself, or required the energy companies to do so—but despite how this 'rule' was been frequently presented by politicians and the media, it was never intended to equate to a guarantee. And policymakers knew this all along because buried on page 106 of the Green Deal consultation was the get-out clause: '*We do not propose to guarantee that the charge will be fully offset by the savings. This would be extremely difficult to either implement or enforce. However, there is nothing to stop organisations going further and guaranteeing that their measures will save customers money*' (DECC 2011).

The moment the Scottish Government read this it should have realised that the energy czars were severely lacking in undergarments but, at least publicly, it seemingly chose to believe that these savings were more than highly optimistic estimates based on flawed data and flawed modelling. Fast forward to 2014 and the figures calculated by the independent expert Chris Goodall pointed to the real savings being a fraction of those claimed by DECC and prompted the Energy Saving Trust to reassess its own figures (Goodall 2014). And just to reiterate, this is something that building scientists have known about for many years, and both DECC and the Scottish Government cannot claim not to have known about it.

Next, we come to the problem of the Green Deal's first financing mechanism, the loans. Prior to its revision to provide grants, Green Deal householders had the option of paying for the measures up front or taking out a 'low' interest loan with one of the Big Six. Needless to say it did not take long for both the left and right wing press to point out the obvious—that borrowing money from an energy supplier against the promise of (inflated) energy savings over a period extending beyond the likely occupation of a property could lead to its devaluation (Bachelor 2013; Poulter 2013). Reading just one of these articles should have been enough to put many householders off signing up to the Green Deal, but those with more than a passing interest in their energy supplier may also have spotted that by restricting the loans to the Big Six they also restricted the choice of supplier for themselves and any future owners of the property. This meant that subscribers to Green Deal loans could not choose to buy their energy from smaller renewable energy suppliers such as Ecotricity and Good Energy who, by the virtue of supplying much more of their energy from renewable sources, provide lower carbon energy at prices that are more immune to increases in the cost of fossil fuels. Not, of course, that those companies would likely want to be associated with such an anti-competitive shambles. This locking in of customers can only have come from a government that believes the idea of a free market only extends as far as those companies with the financial weight to lobby for such concessions, but again the Scottish Government's response was barely a whimper.

The Green Deal and ECO also worked against smaller installers due to the training and registration costs of becoming approved suppliers, which have to be offset against their projected income from the schemes. In the light of the number of measures delivered and the greater capacity of energy companies to manage and recoup their costs through subsidiaries, it is difficult to understand how smaller companies would be able to justify those costs, but easy to understand the financial difficulties they could end up in should the contracts not materialise. Furthermore, when the scheme was changed to provide grants in June 2014, those smaller companies were faced with a sudden glut of contracts they were in a much weaker position to take advantage of. Here again, the Scottish Government cannot claim ignorance as similar criticisms were levelled at its flagship Energy Assistance Package (Baker et al. 2014).

Finally, we move on to a problem that disproportionately affects the Scottish housing stock, that of 'hard to treat' properties. However, before addressing the technical difficulties they pose, it is worth noting that (in theory) greater support should be available for these because many would qualify for funding from both the Green Deal and ECO, but believing this version of reality also requires believing that the level of funding made available for them was ever intended to be proportionally higher for Scotland than for the rest of the UK—presumably through some form of the Barnett Formula adjusted for housing conditions.

7.3 The Problem of Hard to Treat Properties

The term 'hard to treat' covers a wide range of buildings, many of which are more common in the Scottish stock than in the UK as a whole. Depending on how the term is defined they constitute upwards of 25% of the total stock and include all buildings with[1]:

- Solid walls (25% of total stock)
- Tenement flats (23%) and high rise flats (3%)
- Timber frames (5%)
- Flat roofs (4%) and mansard roofs (1%)[2]

Current strategies for refurbishment of building fabric in conventional housing still tend to target the low hanging fruit of installing cavity wall and loft insulation measures in suitable dwellings. These measures offer the most affordable treatments which can reach a significant number of dwellings and deliver significant reductions in energy consumption and emissions and are a key political priority because of their other positive impacts and relative ease of implementation, however, they are often unsuitable for hard to treat buildings.

The design of flat and mansard roofs makes them difficult for retrofitting insulation. Flat roofs are generally maintained by replacement with a more modern cold or warm deck solution, but these are relatively costly and disruptive to occupants. Mansard roofs can be insulated by removing the exterior cladding and installing insulation; however, care needs to be taken to ensure this does not settle, and the exposure of the underlying fabric means that installations must be carried out in periods of dry weather, a particular problem in the Scottish climate (Roaf et al. 2008).

Older timber-framed buildings can be improved using a variety of traditional and more modern methods but are rarely suitable for the off-the-peg solutions generally subsidised by energy efficiency schemes. Modern timber frame buildings tend to be designed with energy efficiency in mind; however, these remain a small proportion of this stock type.

Common reasons for properties not being suitable for cavity wall insulation are more varied but include the following:

- Not constructed with a cavity wall—there is a significant proportion of the current Scottish housing stock constructed with a solid wall structure, ranging from stone tenements to large panel constructions.
- Having a conventional cavity wall construction but also being highly exposed, and therefore considered unsuitable for a full cavity fill (BRE 2002).
- The nature of the cavity wall construction being such that there are concerns over the consequences of a cavity fill. This may be as a result of the state of repair, the quality of the construction, or even the type of construction used.

In other cases, hard to treat buildings may be technically suitable for cavity wall insulation but the difficulties and extra costs involved with non-standard treatment, as well as the potential risks to the building fabric, dictate that other measures are more preferable (Iwaszkiewicz et al. 2010); however, a lack of data means the number of these buildings remains unknown. For properties that cannot be treated with cavity fill insulation safely, other options for adding insulation need to be considered, most commonly internal or external insulation.

Internal wall insulation is possible but usually requires the lining of the structure with a vapour control membrane, and by its nature will require electrical sockets and other service penetrations in the external fabric to be adjusted and sealed. Interstitial condensation risk needs to be carefully controlled, and for small properties, installation can reduce internal dimensions and lead to problems with space standards. Installing internal insulation also requires redecoration and so is best carried out as part of wider home improvements to avoid the problem of occupants having the expectation that the interior will be returned to its original condition and the potential for them to take legal action if this is not achieved (Roaf et al. 2008).

External insulation, such as rendered external wall insulation and insulated over-cladding, offers a solution which covers thermal bridges, reduces dampness penetration risk and can improve the appearance of buildings. However, externally insulating shared buildings such as tenements require the agreement of the households that share them, and is technically difficult for high rise buildings. Shared buildings also require appropriate measures to be selected for and installed in shared areas such as stairwells, as energy losses through walls into these areas can be significant, and (where technically feasible) further savings can be gained by installing draught lobbies.

Both internal and external insulations are also generally most expensive, and so need to be justified by more than the energy savings they provide; however, there is significant potential to improve their cost effectiveness through area-based schemes that target groups of buildings. For example, the cost of erecting scaffolding alone can be prohibitive, not least because of its value on the black market making it an attractive target for thieves, so targeting whole rows of tenements should be an easy way of limiting costs and risks if sufficient finance can be leveraged and residents can be successfully engaged (Baker et al. 2012). A similar argument applies to the benefits of installing packages of measures in a single intervention, and further savings can be made where upgrades to building fabric and systems are designed to complement each other—one example of this being 'deep' retrofitting to the PassivHaus standard (John Gilbert Architects 2013).

A further barrier to installing external insulation is the varied conservation restrictions that remain in many areas of Scotland; however, these generally do not cover the rear facades (backcourts) of tenements, which present a significant opportunity for saving energy as they tend to be poorly constructed compared to the front facades, and often have exposed pipework that contributes to heat losses.

Some parts of the building stock are even more problematic. Significant numbers of non-standard mass-built post-war housing remain across the Central Belt. These require bespoke treatments to ensure that any measures do not have a deleterious effect on the hydrothermal performance of the building fabric and cause air quality, condensation and dampness problems.

This brings us to yet another fundamental problem with the Green Deal, ECO and many other schemes—that of the limited scopes of the measures they support, and this lack of flexibility again works against the Scottish building stock (Baker et al. 2012)—and as such there remains much that could be done to leverage physical improvements, but it would be foolish to believe that these would have been delivered by the Green Deal and ECO, and especially in Scotland. However, before moving on, it should be stressed that in practice 'hard to treat' very rarely means technically difficult to treat. The real difficulties lie in tackling the socio-economic conditions that limit householders' abilities to improve their homes (Roaf et al. 2008), particularly those in mixed-tenure multiple occupancy buildings and the conventional social science-led approaches to policy making that cannot sufficiently address such complex problems (Castellani 2014; Maiden et al. 2016).

7.4 Getting it Right on Energy Efficiency

It can be incredibly frustrating to observe a series of governments get things so badly wrong on energy efficiency when other countries have managed to do so much better, and whilst some of these successes have as much to do with long-standing cultural attitudes than clever political thinking, it should not be beyond the abilities of policymakers to distinguish between what should and should not work in Scotland, be that for technical, environmental, social, economic or cultural reasons, or because they are being sold snake oil.

Of those policies which have been proven to work probably the most successful, and appropriate for the UK, is the Residential Energy Conservation Ordinance (RECO) which began life in Berkeley, California, and has since been adopted across the more forward-thinking states of the USA. Under RECO, any home being sold, changing tenancy or being extended must be brought up to the latest building standards within a year of the sale, change of new tenancy or extension. The clock starts ticking from the point of sale, transfer of tenancy, or a significant extension to the property, and does not reset if the home is resold within the year. The key benefit of this is that, as studies have shown, householders are most receptive to energy upgrades at transition points in their lives. Another benefit is that it serves to level the playing field as the costs of retrofitting energy-inefficient properties have to be factored into the market value. RECO has been hugely successful and cost-effective but has had years to bed in and take effect, so whilst it could have similar results in the UK, other approaches are still needed if the building stock is to be upgraded in time to meet its emissions-reduction targets (Emmanuel and Baker 2012). Someone at DECC clearly was aware of RECO, because you can see its influence on some of the finer details of the Green Deal, but sadly none of its teeth.

Another lesson the UK could learn from other countries is how to leverage packages of measures drawn from a much wider scope of options. Experience from the Leadership in Energy and Environmental Design (LEED) rating systems used in the USA and the building standards of the Netherlands and others shows that this is best achieved through requiring a combination of proscribed basic measures and 'allowable solutions' to meet a minimum improvement in performance (as per RECO). However, this of course requires expert input to ensure the scope of the proscribed measures is appropriate to capture all building types (Emmanuel and Baker 2012).

Another tool that could be employed is to use council tax as a lever for improvements, with rebates available on proof of installation. Any such charges would have to be carefully introduced to be publicly and politically acceptable and would not capture most of the woefully poor private rented sector without landlords being made responsible for paying council tax, but the idea now has a precedent in Scotland following a recent pilot by Highland Council (Highland Council 2016).

An even blunter option would be to introduce exponential pricing brackets for energy use. This system has the benefit of capturing profligate energy users, and the profiling of individual energy consumption made possible by smart meters means brackets could be individualised to different household types, but again changes to tenancy law would be needed to ensure that tenants do not bear the costs for absentee landlords unwilling to pay for energy efficiency upgrades. Safeguards would also be needed to make sure those households who lack the education or ability to control their own energy consumption receive support to help them reduce their bills, along with needs-based benefits for vulnerable households, such as those with long-term health problems that affect their energy use. Therefore, whilst this might seem an appealing option for a government with an unfettered belief in the free market, it also means that any attempt to introduce it by such a government should be treated with a great deal of caution.

Nevertheless, there is evidence to suggest that focusing on 'fuel-hungry' households, if done fairly, is effective. For example, lower occupancy households tend to be more energy hungry but are also more responsive to efficiency interventions, whilst interventions to households who use significantly less than average amounts of energy will yield proportionally smaller savings (Lomas 2010). It is also easier to identify and target profligate energy users, at least where that profligacy relates to ability to pay rather than inability to manage household finances, because the key indicators of proportionately higher energy consumption are simpler and easier to incorporate in models and policies (Jones and Lomas 2015).

7.5 Changing Behaviour

The energy efficiency and behaviour change policies developed by both Westminster and Holyrood largely target three key sectors: homes and communities; business, industry and the public sector; and transport. However, the inherent complexities of human behaviour mean it is difficult to attribute emissions reductions to specific policy levers or behaviour changes, and the term behaviour change is itself problematic if it is defined, as it is common in Scottish Government publications, as also including 'one off' behaviours such as installing insulation. Nevertheless, much can still be inferred from existing targets and aspirations.

The Scottish Government's policies targeting homes and communities, many of which incorporate at least some element of behaviour change, are expected to yield a 14% reduction on 1990 levels by 2020; those targeting business, industry and the public sector are expected to contribute a further 14%, and transport bears the brunt of contributing 27%. None of these targets are likely to be easy to achieve. The reduction from homes and communities is equal to 38% of the 1990 baseline, proposals for business and industry are sensitive and subject to the economic recovery, and the projections for transport acknowledge that emissions are still increasing and the impact of policies may be based on 'optimistic assumptions' (Audit Scotland 2011; Scottish Government 2013).

As an aside, whilst this chapter is focused on discussing behaviour change in the context of the built environment, it is important to remember that much can be achieved, and indeed much is at stake, from changing transport behaviours. Here again, there are problems of political palatability, for example reducing air travel and minimising freight and distribution networks, but these also present opportunities in a culture that is naturally receptive to local sourcing and nature tourism. There is also the need to increase renewable generation to meet the new demand from electric vehicles, and more urgently the need to install new infrastructure to enable them to be a real alternative to the internal combustion engine, as well as upgrading and expanding public transportation networks, and the need to upgrade telecommunications infrastructure to enable more people to work remotely.

7.6 Achieving Behaviour Change

Effectively enabling behaviour change is recognised by government departments and government-funded organisations as an essential component of meeting the emissions-reduction targets set by the UK and Scottish Climate Change Acts, for example in sustainable development (Defra 2008; SDC 2010), energy (Carbon Trust 2015), transport (DfT 2010) and society (Scottish Government 2009). However, this also demonstrates the complexity of the task and requires a strategy that goes significantly further and is more nuanced than the policy initiatives that have sought to target specific public behavioural changes in the past, but for which the Scottish Government and its partners can claim some achievements.

Examples of successful behaviour change strategies include drink-driving (Mann et al. 2001), reduced speeding (Pilkington and Kinra 2005), use of seat belts (Jochelson 2007) and smoking (Adshead and Thorpe 2007), but in each of these cases it is useful to note the length of time each took to bring about behaviour change and the extent to which each behaviour has become pervasive amongst the UK population. What such successes tend to have in common is that the changes are easily communicable, and target single habits with simple, tangible messages about the benefits and costs of making the change and are often legally enforceable. In contrast, the benefits of energy efficiency behaviours are deferred and harder to relate to. Habitually not wearing a seat belt carries a high risk of an on-the-spot fine (as well as points on a driving license), whereas habitually switching off lights may save a similar amount of money but those savings are deferred to a number on an energy bill that needs to be related to previous figures.

However, it is arguable that classifying single actions such as upgrading insulation as 'behaviour changes' muddies the water as these can be prompted directly through regulation and incentivisation (Baker et al. 2012). To gain a better handle on the more complex task of what many would understand by 'behaviour change'—long-term habitual and lifestyle changes—it may be better to treat all savings from single, practical actions (such as installing insulation) as purely savings from technical solutions, and to weight the numbers of installations of different measures against the capacity of households to install them.

This leaves us with a set of behaviours that vary in frequency from many times a day (e.g. switching off unused lights and appliances) to annually (e.g. avoiding flying when taking a holiday), and that vary in convenience from easy (e.g. remembering not to over-fill a kettle) to requiring substantial lifestyle adjustment (e.g. switching regular travel away from the private car). As with all habitual changes, they initially require an element of prior planning but can become 'automated' behaviours if repeated regularly enough (Lally et al. 2010). This model of emphasising early repetition to achieve the normalisation of a behaviour is an established idea in psychology and one which presents another problem for changing energy behaviour. Whereas Lally's study measured three habit changes (eating, drinking and exercising) for which it was assumed (correctly) that self-motivation and willpower would be enough to bring about automation, the same can not be assumed for most energy efficiency behaviours—aside from the minority for whom the environment is a strong enough motivation in itself.

This means that frequent reinforcement should be critical to changing energy behaviours, but it begs a number of important questions. How do you reinforce such changes to behaviours that occur with high frequencies in the privacy of a home? How do you reinforce travel behaviour changes when the roads are still flooded with private cars and access to cheap air travel has become a societal norm? And, ultimately, what authority can be claimed by politicians and policymakers when (for example) public buildings remain lit up at night, and the carbon cost of 'Earth Summits' can be greater than the annual emissions from entire countries?

One proposed solution is to tackle the supply side of the problem by carbon rationing, for example by Personal Carbon Allowances (PCAs, for example, White et al. 2013), but whilst such an idea is a highly rational solution that could also leverage greater societal equality it is easy to imagine the fortunes of any political party brave enough to include it in their manifesto. Compare this to the conventional but far less-effective method of reinforcement through advertising campaigns, and how much they would need to be scaled up to prompt the correction of common energy inefficient behaviours sufficiently frequently to normalise alternative behaviours over an entire population, and you begin to realise size of the gulf between political aspirations and reality. So, in the light of this, it should come as no surprise that policymakers are left grasping at anything that could influence energy behaviours using that most basic of motivators—cost—using a populist lever—consumer demand.

7.7 Smart Meters

The UK's latest great hope, it seems, is smart metering, and there are many positive things that could come of it. Smart meters should one day enable households and businesses to have unparalleled control over their energy use, and the prices they pay for it. Early incarnations of this technology are already in use by larger businesses, and domestic technologies are now coming on to the market, but to really take off they will need to persuade consumers that the functionality they offer and the money they save is worth it—and that cost will be measured in information as well as pricing.

Many countries around the world are now engaged in programmes to facilitate the mass roll-out of smart meters, including Sweden, Italy, the USA, Canada, Australia and New Zealand (Germany having curtailed its own programme). In the EU, the Energy End-Use Efficiency and Energy Services Directive (2006/32/EC) (also known as the Energy Services Directive) requires the installation of basic (real-time display only) meters in all new buildings, and when existing meters are replaced. However, the more recent Directive 2009/72/EC (Concerning the rules for the internal market in electricity) downgrades this to '*where economically reasonable and cost-effective*'.

The term 'smart meter' covers an increasingly wide range of devices used to inform the occupants of a building about their energy consumption and regularly transmit that data to utility companies, and with a UK-wide roll-out aiming to have them installed in all homes by 2020 they have become very popular with politicians and policymakers. In theory, the roll-out will pump prime the market so that smart meters become a normalised, desirable, technology, and demand becomes self-sustaining.

Smart meters can also be used for monitoring gas and water consumption, and more advanced devices can be used to remotely control building services and appliances. The term is also frequently used to include those devices that display consumption data but do not broadcast it beyond the building, which may be justifiable for those meters capable of measuring and disaggregating the energy consumption attributable to individual building services or appliances. The more basic earlier devices that do not transmit data beyond the home and are more correctly termed 'energy consumption indicators' (ECIs).

Providing real-time consumption data to occupants has been found to be effective in reducing energy consumption when used either in place of or in conjunction with other behavioural levers (Faruqui et al. 2010; Wood and Newborough 2007). Basic smart meters usually display the amount of energy being consumed, the cost (financially and in CO_2), and how this compares to previous consumption—for example that of the previous day or a monthly average. These meters rely on occupants learning about their energy consumption by switching services/appliances on and off and noting the changes, but more advanced devices (e.g. those that can be linked to home computers) can provide a greater range of data outputs. The way that data is displayed is also important most consumers would struggle to relate to figures for CO_2 savings so cost will almost certainly be the dominant metric, but whereas those figures should be meaningful enough for businesses there may be more work to be done to make them meaningful for the average household. So, whilst it remains to be seen which models will become dominant in the roll-out, policies to encourage their take-up need to be sensitive to enabling more energy-aware households to take advantage of more sophisticated devices, whilst also enabling vulnerable and information-poor households to understand the information provided by them, or to install models that enable remote management by a third party.

At the moment smart meters have two main uses: to provide actual consumption data to utility companies to allow them to issue accurate bills (rather than estimated bills, which are frequently contested), and to inform occupants about how much energy they are using and what is responsible for this. However, more advanced devices which offer remote control over heating and major appliances are already on the domestic market, and even greater functionality is available for those companies who can afford the investment. Four measures that can already be combined with smart meters are programmable thermostats; zoned heating and cooling; remote control of HVAC systems; and outlet-level appliance monitors that can automatically disconnect appliances to eliminate leakage currents (Meyers et al. 2010).

Another key argument in favour of smart meters is that enabling building occupants to have greater control over their energy (and water) consumption leads to greater *optimisation* of energy use (for example, Baker and Rylatt 2008; Willis et al. 2010). This is a huge benefit to building energy modellers as it means buildings are more likely to perform as predicted by simpler models—and even more complex models invariably assume stronger relationships between energy use and a small number of well-studied factors such as floor area, occupancy and the type of heating system.

Although market penetration is still relatively low and studies of the impact of basic meters have not found drastic reductions in consumption, e.g. around 7% in the USA (Ueno et al. 2005), this is far from a reason to dismiss their potential. If the demand for smart meters is to really take off the direction of their evolution, and the way they are marketed, should focus as much on being able to remotely turn appliances on as to turn them off. As discussed previously, investing in a technology to save money on (future) energy bills may be a hard sell, but selling a technology that enables householders to come home to a warm home and a hot meal ready in the oven should be a different game entirely. This greater level of control can be achieved actively via smartphone apps, and from here it is just a simple step to enable passive control by using location data and geo-fencing to automatically control systems and appliances according to where the occupants are—so if the smartphone finds itself connected to a network in another country it can assume its owner would not be returning that night and switch off anything that has been left on and throttle down 'always on' appliances such as fridges. However, as yet the benefits of smart meters lag far behind the potential.

7.8 Smart Meters—The Reality

The UK Government has a dismal record on delivering IT projects, and the smart meter roll-out is following the same course. At the heart of this looming failure is another basic error in understanding how to build a network, but first we need to understand how the system will work.

The current roll-out will see households equipped with a wireless hub (provided by one of two suppliers) that will collect data from the (separate) electricity and gas meters (supplied by third parties) and transmit it via a home hub to the new Data Communications Company (DCC), which will then pass it on to the energy suppliers. The complexity of this system, which includes least two different hub designs and an unlimited number of meter designs means DECC had to adopt its own wireless standard for the systems. In its wisdom, it chose to reject both the Wi-Fi and Bluetooth standards used by almost all consumer telecommunications devices and opted for ZigBee, a near-defunct system largely confined to highly niche applications, but whose Smart Energy Profile (SEP) standard has been adopted by some US utility companies. This means that although smart meter manufacturers may build in Wi-Fi or Bluetooth functionality into their devices, for example to enable connection to smartphones, those channels cannot be used to connect directly to the hub and on to the DCC. It is also worth noting that ZigBee operates at 2.4 GHz, and so may interfere with existing Wi-Fi and Bluetooth networks, and the low power transmitters it uses to communicate between the fixed meter and the hub mean they will have to be placed so close together that they may interfere with each other. The low power is possible because ZigBee is a 'mesh' network which saves energy by forming and sharing a network with other devices within range—assuming of course there are other devices in range. This could still make sense with a sufficient lead-in time to build capacity amongst developers and get sufficient numbers of hubs into homes, but revision 1.3 of the SEP standard, which was needed to meet new requirements added by DECC, has left smart meter manufacturers just a few years to upgrade and test their devices (assuming of course this will be the final revision).

Furthermore, whilst the smart meters will be able to display precise measurements of energy consumption, they will not be able to display precise costs, because although they are likely to be user-adjustable for different tariffs, they will be unlikely to calculate the additional (and frequent) adjustments made by energy suppliers, such as dual fuel savings and other discounts. As such, this information will have to be given with the caveat that it is only indicative (Hunn 2013).

All this is just a summary of the litany of technical issues that have been brought about by the political desire to meet the demands of stakeholders (i.e. the energy companies) rather than meeting the needs of consumers through by designing the system around simplicity and interoperability. Those demands have arisen because, unlike many IT companies, the internal data management infrastructure and systems used by the energy companies are archaic and not designed for the demands of managing and processing the comparably huge volumes of data smart meters will generate. Furthermore, these companies tend to lack the knowledge and expertise for developing large-scale communications networks that could have been brought by a specialist provider, and the costs involved in developing and implementing that expertise mean they have little incentive to go beyond the minimum requirements. As things stand, the system will be designed around the demand that it works with all those internal systems, rather than requiring energy suppliers to update them to meet a common standard.

If, as it should have done, DECC had required the system to be designed around what it needs to do, and may need to do in future, then we would be looking at a radically different roll-out. A much more effective way of designing the network would have been to put a standard system under the control of the distributors, who already occupy this space in the physical chain between suppliers and consumers. It could also have done without the additional complexity of requiring smart meters for gas (and therefore the need for hubs) as this consumption is much less sensitive to human factors (EEA 2013)—and devices already on the market (and not using ZigBee) could still be converged to allow remote control of heating systems.

The meters themselves would also have been much simpler, and therefore cheaper, with the bulk of the processing done on central servers, and the in-home meters being largely displayed devices. This would mean that changes to tariffs and other aspects of the system could be applied in real time and would minimise the frequency of firmware upgrades to the meters themselves. This is a lesson in cloud computing that IT companies learned a long time ago and, armed with this significant advantage, it seems unlikely that the technology giants will stand by and see the profits to be made from managing massive amounts of data handed to the energy companies. Google signalled its intentions to contest the market by buying up Nest, whose smart thermostats and smoke alarms just happen to use ZigBee (Carroll 2014). And if Google (or now Alphabet) is getting into energy management, then why would one of the world's most profitable companies stop there? It has already amassed investments of over $1 billion in renewables, which it describes as making '*attractive*' financial returns (Google 2014), and Apple has also joined the game (MDN 2013). So, we could one day be buying green energy from a technology giant, but how would consumers feel about handing even more of their data to one company, even one that claims to 'do no evil'?

7.9 Conclusion: Changing Attitudes

Consumers are becoming increasingly aware of how their personal is collected and used, and also of the value of that data. A glimpse of quite how valuable this personal data is to a major IT company was given away by Facebook's purchase of the instant messaging firm WhatsApp, from which analysts were able to derive an average value of $140 for a personal Facebook page (Kuchler 2014). However, whereas Facebook's services are largely restricted to social media, this value would be expected to be greater for a more diversified company such as Google, Apple or Microsoft. Now imagine the value of one account that encompasses a customer's energy use as well as all that other personal data.

What Google has achieved goes way beyond becoming the world's number one search engine and free email client (both of which were already highly contested markets when it entered them); it has fundamentally changed attitudes to data privacy by offering services that consumers have deemed valuable enough to be worth handing over increasing amounts of it and in a very small amount of time. Doing this with energy data, particularly if the data manager is also the energy supplier, could revolutionise the energy market, and if that revolution were to drive a change in attitudes to energy use, then the potential benefits could be immense—but how much more data will consumers be happy to give away?

The revelations concerning the collection of personal data by US's National Security Agency and the UK's GCHQ may be signs of how far the public are prepared to be led down this road, and it is far from inconceivable that this could affect the Government's plans for smart meters—especially as the UK's energy companies are distrusted by the majority of the public (Pickard and Chazan 2013). In addition, the hacking and release of personal data such as credit card details from many major data managers have begun to embed a healthy distrust of the safety of that data, even when held on 'secure' systems. The UK Government has collected individual energy consumption data for many years (Baker 2007), but this is not something that the public have generally been aware of or concerned about. However, with the greater detail provided by smart meters and the national roll-out will come to a greater awareness of what this data could be used for and its value to energy companies and governments, and to those who would seek to steal it from them, so a positive public reaction is now far from guaranteed.

As previously discussed, the benefits of all this are being sold largely on the financial savings, and in a rational society, this should be all that is needed to sell the technology and achieve the holy grail of changing attitudes to a problem that is often far from the public consciousness. But this growing distrust, along with a poor technology strategy and a lack of commitment and transparency over how this data will be managed, could throw a big spanner in the works no matter how hard the Government and its partners work to address public concerns—or even if that market is eventually ceded to more trusted data managers.

As things stand, the future for smart meters looks far from certain. Although they can still be expected to play an increasingly important role in managing and reducing our energy consumption, the challenge ahead is, more than ever, about achieving attitudinal change. And the hardest part of that will be managing these changes in a way that addresses the biggest problem of all—that humans are innately human.

Notes

1. They also include an estimated minimum of 4121 park and residential mobile homes, which are classed as products and so fall under the British Standards rather than the Scottish Building Regulations.
2. Figures are for 2008 but have varied little over time. More recent data for some classifications can be found in: Scottish Government (2014).

References

Adshead, F., and A. Thorpe. 2007. The Role of the Government in Public Health: A National Perspective. *Public Health*, 121 (11): 835–839.

Audit Scotland. 2011. *Reducing Scottish Greenhouse Gas Emissions* (online). http://www.audit-scotland.gov.uk/docs/central/2011/nr_111208_greenhouse_gases.pdf.

Bachelor. 2013. *Why the Green Deal Could Make Your Home a Hard Sell.* The Guardian, 27 Jan 2013 (online). http://www.theguardian.com/money/2013/jan/27/green-deal-home-hard-sell.

Baker, K.J. 2007. Sustainable Cities: Determining Indicators of Domestic Energy Consumption. PhD thesis. Institute of Energy and Sustainable Development, De Montfort University, Leicester, UK.

Baker, K.J., R. Emmanuel, and M. Phillipson. 2012. *Support for RPP2—Housing Futures.* Report for ClimateXChange Scotland (online). http://www.scotland.gov.uk/Resource/0038/00389071.pdf.

Baker, K.J., R. Emmanuel, and M. Phillipson. 2014. Review of the Energy Assistance Package. Report for the Scottish Government.

Baker, K.J., and M. Rylatt. 2008. Improving the Prediction of UK Domestic Energy Demand Using Annual Consumption Data. *Applied Energy* 85: 475–482.

BRE. 2002. Thermal Insulation: Avoiding Risks. BRE Report 262, Building Research Establishment, Watford, UK.

Carbon Trust. 2015. *Low carbon behaviour change: The £300 million opportunity* (online). http://www.carbontrust.com/resources/reports/advice/low-carbon-behaviour-change/.

Carroll, R. 2014. *Google buys Nest Labs for $3.2 bn in bid for Smart Home-devices Market.* The Guardian, 14 Jan 2014 (online). http://www.theguardian.com/technology/2014/jan/13/google-nest-labs-3bn-bid-smart-home-devices-market.

Castellani, B. 2014. Complexity and the Failure of Quantitative Social Science. Discover Society Nov 2014.

DECC. 2011. *The Green Deal and Energy Company Obligation—Consultation Document.* Department for Energy and Climate Change, London (online). http://www.decc.gov.uk/assets/decc/11/consultation/green-deal/3607-green-deal-energy-company-ob-cons.pdf.

DECC. 2015. *Green Deal and Energy Company Obligation (ECO): Monthly statistics.* Department for Energy and Climate Change, London (online). https://www.gov.uk/government/statistics/green-deal-and-energy-company-obligation-eco-monthly-statistics-august-2015.

Defra. 2008. *Changing Behaviour* (online). http://www.defra.gov.uk/sustainable/government/what/priority/changing-behaviour.htm.

DfT. 2010. *Behaviour Change: What Works for Transport? Think Piece project* (online). http://www.dft.gov.uk/pgr/scienceresearch/social/behaviour-changes/.

EEA. 2013. *Achieving Energy Efficiency Through Behaviour Change. European Environment Agency* (online). http://www.eea.europa.eu/publications/achieving-energy-efficiency-through-behaviour.

Emmanuel, R., and K.J. Baker. 2012. *Carbon Management in the Built Environment*. London: Routledge.

Faruqui, A., S. Sergici, and A. Sharif. 2010. The Impact of Informational Feedback on Energy Consumption: A Survey of the Experimental Evidence. *Energy* 35: 1598–1608.

Goodall, C. 2014. *Actual Energy Savings From Efficiency Measures Only Half of What is Officially Claimed*. Carbon Commentary (online). http://www.carboncommentary.com/2014/01/18/3368.

Google. 2014. *Investing in a Clean Energy Future* (online). http://www.google.co.uk/green/energy/investments/.

Highland Council. 2016. Council Tax—Discounts For Properties (online). http://www.highland.gov.uk/info/701/council_tax/407/council_tax_-_discounts_for_properties/8.

Hunn, N. 2013. *Smart Metering is Fcuked*. WiFore Consulting (online). http://www.nickhunn.com/wp-content/uploads/downloads/2013/11/Smart-Metering-is-FCUKED.pdf.

ICStart. 2013. The Green Deal and Sustainable Refurbishment of Traditional Buildings. In *Conference held at Glasgow Caledonian University on 29 Feb 2012* (online). http://www.cicstart.org/events/events/212,481/Conference–The-Green-Deal-and-Sustainable-Refurbishment-of-Traditional-Buildings.html.

Iwaszkiewicz, C., J. Christofides, W. Wright, R. Thompson, J. Connaughton, B. Heyes-Lewin, and R. Newey. 2010. *Study on Hard to Fill Cavity Walls in Domestic Dwellings in Great Britain*. Report to DECC, Ref: CESA EE0211 http://www.decc.gov.uk/assets/decc/what%20we%20do/supporting%20consumers/saving_energy/analysis/788-hard-to-fill-cavity-walls-domestic.pdf.

Jackson, T. 2009. *Prosperity Without growth*. Sustainable Development Commission, UK (online). http://www.sd-commission.org.uk/data/files/publications/prosperity_without_growth_report.pdf.

Jochelson, K. 2007. The Role of the Government in Public Health: A National Perspective. *Public Health* 121 (11): 1149–1155.

John Gilbert Architects. 2013. *PassivTEN: Upgrading Glasgow's Tenements to PassivHaus standard*. John Gilbert Architects, Glasgow (online). http://www.johngilbert.co.uk/files/130128RPT_PassivTEN_NoAppendix_v4_web.pdf.

Jones, R.V., and K.J. Lomas. 2015. Determinants of High Electrical Energy Demand in UK Homes: Socio-economic and Dwelling Characteristics. *Energy and Buildings* 101: 24–34.

Kuchler, H. 2014. *WhatsApp 'Sticker Shock' Purchase Offers Facebook Mobile Leverage*. Financial Times, 20 Feb 2014 (online). http://www.ft.com/cms/s/0/abd4a734-99d9-11e3-b3a2-00144feab7de.html#axzz2txqFpntC.

Lally P., H.M. Cornelia, V. Jaarsveld, H.W.W. Potts, and J. Wardle. 2010. How are Habits Formed: Modelling Habit Formation in The Real World. *European Journal of Social Psychology* 40 (6): 998–1009.

Lomas, K.J. 2010. Carbon Reduction in Existing Buildings: A Transdisciplinary Approach. *Building Research & Information* 38: 1–11.

Maiden, T., K.J. Baker, and A. Faulk. 2016. *Taking the Temperature: A Review of Energy Efficiency and Fuel Poverty Schemes in Scotland.* Report for Citizens Advice Scotland by CAG Consultants, Glasgow Caledonian University, Andrew Faulk and the Energy Agency.

Mann, R.E., S. MacDonald, G. Stoduto, S. Bondy, B. Jonah, and A. Shaikh. 2001. The Effects of Introducing or Lowering Legal per se Blood Alcohol Limits For Driving: An International Review. *Accident Analysis & Prevention* 33 (5): 569–583.

MDN. 2013. *Apple's Ground-Breaking Bet on Its Energy Infrastructure.* Mac Daily News, 18 Nov 2013 (online). http://macdailynews.com/2013/11/18/apples-ground-breaking-bet-on-its-clean-energy-infrastructure-with-video/.

Meyers, R.J., E.D. Williams, and H.S. Matthews. 2010. Scoping the Potential of Monitoring and Control Technologies to Reduce Energy Use in Homes. *Energy and Buildings* 42: 563–569.

Pickard, J., and G. Chazan. 2013. *Energy Companies Feel The Heat Amid Anger Over Price Increases.* Financial Times, 12 Nov 2013 (online). http://www.ft.com/cms/s/0/7c23cf66-4bbd-11e3-a02f-00144feabdc0.html#axzz2txqFpntC.

Pilkington, P., and S. Kinra. 2005. Effectiveness of Speed Cameras in Preventing Road Collisions and Related Casualties: A Systematic Review. *British Medical Journal* 330: 331–334.

Poulter, S. 2013. *The 25 year 'Green Deal' Loans That Could Prevent You Selling Your House.* The Daily Mail, 14 Sep 2013 (online). http://www.dailymail.co.uk/news/article-2420496/The-25-year-green-deal-loans-prevent-selling-house.html.

Roaf, S., K.J. Baker, and A. Peacock. 2008. *Evidence on Hard to Treat Properties.* Scottish Government publication.

Scottish Government. 2009. *SEABS 08: The Scottish Environmental Attitudes and Behaviours Survey 2008.* Scottish Government Social Research, Edinburgh (online). http://www.scotland.gov.uk/Publications/2009/08/03100422/0.

Scottish Government. 2013. *Low Carbon Scotland: Meeting our Emissions Reduction Targets 2013–2027.* The Second Report on Proposals and Policies (RPP2) (online). http://www.scotland.gov.uk/Publications/2013/06/6387/0.

Scottish Government. 2014. *Scottish House Conditions Survey 2013—Key Findings* (online). http://www.gov.scot/Resource/0046/00465627.pdf.

Scottish Government. 2015. *Scottish House Condition Survey 2014.* Scottish Government publication.

SDC. 2010. *The Future is Local.* Sustainable Development Commission, July 2010 (online). http://www.sd-commission.org.uk/publications.php?id=1093.

Ueno, T., F. Sano, O. Saeki, and K. Tsuji. 2005. Effectiveness of an Energy-Consumption Information System on Energy Savings in Residential Houses Based On Monitored Data. *Applied Energy* 83: 166–183.

White, V., J. Thumim, and I. Preston. 2013. *Personal Carbon Allowances: Distributional implications associated with personal travel and opportunities to reduce household emissions.* Centre for Sustainable Energy, Bristol, UK (online). http://www.cse.org.uk/downloads/file/project_paper_3_personal_carbon_allowances.pdf.

Willis, R.M., R.A. Stewart, K. Panuwatwanich, S. Jones, and A. Kyriakides. 2010. Alarming Visual Display Monitors Affecting Shower End Use Water and Energy Conservation in Australian Residential Households. *Resources, Conservation and Recycling* 54: 1117–1127.

Wood, G., and M. Newborough. 2007. Energy-use Information Transfer for Intelligent Homes: Enabling Energy Conservation with Central and Local Displays. *Energy and Buildings* 39: 495–503.

Author Biography

Dr. Keith Baker is a Researcher at the School of Engineering and the Built Environment, Glasgow Caledonian University and a co-founder of the Initiative for Carbon Accounting (ICARB). He specialises in energy in the built environment, with a particular focus on fuel poverty, energy efficiency and sustainable building materials.

CHAPTER 8

Crossing the Rubicon: The 2015 Renewable Electricity Reforms and Implications for Scotland

Anna L. Berka, Jelte Harnmeijer and Bill Slee

8.1 Introduction

On 18 June 2015, the UK put in place a number of policy amendments signalling an intention to phase out most technology-specific support schemes for low carbon energy. The reforms came 26 years after the introduction of the UK's first renewable energy support scheme. In spite of the Conservative Party manifesto commitments to reduce support for some types of renewables, the speed and extent of the proposed changes caught industry analysts by surprise. Within months following the UK General Election in May 2015, the sector saw the abandonment of at least 23 large-scale projects representing around 2.7 GW, including one of two carbon capture and storage projects, with crowdfunded loan providers such as the Trillion Fund halting renewable energy loans and public loan providers across the UK halting ongoing loan negotiations.

A.L. Berka (✉)
Department of Forest Sciences, University of Helsinki, Helsinki, Finland
e-mail: anna.h@scene.community

J. Harnmeijer
University of Edinburgh, Edinburgh, Scotland, UK
e-mail: Jelte.Harnmeijer@hutton.ac.uk

J. Harnmeijer · B. Slee
James Hutton Institute, Dundee, Scotland, UK

G. Wood and K. Baker (eds.), *A Critical Review of Scottish Renewable and Low Carbon Energy Policy*, Energy, Climate and the Environment,
DOI 10.1007/978-3-319-56898-0_8

DECC stated that 7 GW of onshore wind projects in planning were likely to miss the early April 2016 deadline for the expiration of the Renewable Obligation and risked being stranded, adding uncertainty as to how the UK would meet its 2020 targets (EU Commission 2015). RenewableUK threatened legal action, drawing on a clause under the Levy Control Framework (LCF) that stated that the government would '***not make retroactive changes to support levels to maintain investor confidence***'. Within months, the UK dropped out of the top ten countries in the Renewable Energy Country Attractiveness Index (RECAI) (Ernst and Young 2015).

The policy shock has been sufficient to draw comments from senior commentators such as Professor Jacqueline McGlade, chief scientist of the United Nation's (UN) environment programme, who argued: '***What's disappointing is when we see countries such as the United Kingdom that have really been in the lead in terms of getting their renewable energy up and going—we see subsidies being withdrawn and the fossil fuel industry being enhanced***'. However, the reforms mirrored ongoing policy shocks in Spain, Italy, Germany and Denmark—all global leaders in renewable energy that have scaled back on renewable energy support as a result of a resurgent discourse around short-term competitiveness and consumer protection (Lauber and Jacobsson 2016). In Germany and Denmark alike, Conservative Party majorities elected in the aftermath of the financial recession provided an opportunity for dormant but long-standing opposition to public support for renewable energy to manifest itself in the form of annual caps and steeper subsidy degressions (Lauber and Jacobsson 2016). These events highlight the formidable challenge of the renewable energy transition in reforming reigning structural, technical and market formations, and resulting political struggles that emerge when renewable energy deployment reaches 15–40% of total electricity supply, where it begins to challenge incumbent utilities and where fundamental market restructuring becomes necessary (Klessmann et al. 2008).

In what follows, we outline the 2015 UK policy reforms and discuss the likely consequences for Scotland, focusing on renewable electricity policy. We place these events in a historical and international context, elaborate on the politics of the affordability and subsidisation of renewable energy and analyse these reforms in the context of emerging renewable energy innovation systems. Drawing on similar policy shocks in Spain, Australia and Germany, we discuss the likely implications of the 2015 policy reforms for the renewable energy industry in Scotland. Specifically, we ask whether the Scottish Government has the capacity to provide public support for a sector which it has promoted so actively and review the prospects for 'subsidy-free renewable energy'.

8.2 An Overview of 2015 UK Policy Reforms

Three key support mechanisms currently support renewable electricity generation by independent power producers in the UK: the Renewables Obligation (RO), the Feed-in Tariff (FiT) and the Contracts for Difference Feed-in Tariff (CfD FiT). Between them, this triad of mutually exclusive mechanisms spans the breadth of major renewable energy technology types and capacity classes. Although there are major differences in how they operate, all three subsidise renewable energy output over time. As a result, expectations about their future form and function play a major role in building—or undermining—investor confidence. The following outlines the policy amendments introduced in 2015 for each scheme.

8.2.1 Renewables Obligation

Replacing the nine year old Non-Fossil Fuel Obligation (NFFO), the RO was introduced in England, Wales and Scotland in April 2002, and in Northern Ireland three years later. Unlike the NFFO, which was originally conceived to support nuclear power, the RO was specifically designed to support renewable electricity generation (Mitchell and Connor 2004). The RO operates as a green certification mechanism and is a classic market-based technology performance standard. RO certificates (ROCs) are sold to electricity suppliers to meet their annual incremental quota for renewable electricity delivery, with annual targets set by the Scottish Government (Mackenzie 2009). Suppliers that do not meet targets pay into a buy-out fund at a rate that is pegged to the Retail Price Index (OFGEM 2011). The buy-out fund is annually redistributed amongst suppliers according to the number of ROCs submitted, rewarding ROC acquisition below supplier obligation levels and reducing consumer electricity cost where supply companies submit relatively large numbers of ROCs. Because ROCs are traded at market price, their value fluctuates annually depending on quota levels in relation to the number of ROCs available on the market. While the advantage of the RO is that it allows control over expenditures, the disadvantage of the RO is that generators are exposed to the political and commercial risk of changing annual targets and ROC prices.

The RO has seen several revisions since 2002, most notably the introduction of technology-specific banding in 2009 which was implemented in order to increase the competitiveness of less mature technologies. With few exceptions, any 'large-scale' (>5 MW) installation commissioned after 2002 and before the recent CfD FiT auction under the EMR has benefitted from the sale of certificates. The RO also supports virtually all renewables projects (of all scales) in Northern Ireland, where small-scale FiTs are not available. As a result, the RO support mechanism has underpinned the majority of renewables capacity developed in the UK and has been the most important mechanism for progress towards achieving the UK's national 2020 target of 15% of total energy to come from renewable sources.

The RO scheme was envisaged to remain open until 31 March 2017, when it was to be fully replaced by the CfD scheme. However, citing unexpectedly rapid uptake of solar PV, the scheme was closed early to 'large-scale' (>5 MW) solar on 1 April 2015 (DECC 2014a). On 18 June 2015, Secretary of State for Energy and Climate Change Amber Rudd announced that the RO scheme would close a year earlier (on 1 April 2016) for onshore wind (Rudd 2015a). At the time, it was left unclear as to under which circumstances onshore wind projects would be eligible for an RO 'grace period', initiating a chaotic back-and-forth dialogue between government and industry bodies.

18 June 2015 marked the beginning of a period of heightened uncertainty for the UK wind industry, which was soon to spread to other generation technologies less exposed to critique on the grounds of cumulative landscape visual impact. A month later, on 22 July 2015, DECC published a consultation on closing RO support for solar < 5 MW as of 1 April 2016 (DECC 2015a). This consultation also raised the spectre of English and Welsh solar projects accredited after 22 July not receiving RO grandfathering rights, greatly diminishing expected project value and heightening risk. The decision associated with this consultation, which closed on 2 September 2015, was still pending at the time of writing, but few industry insiders expect significant modifications of the changes proposed. In Northern Ireland, meanwhile, the Department of Enterprise, Trade and Investment (DETI) launched a two-week consultation on 30 September to close the Northern Ireland RO for new onshore wind projects on 1 April 2016, in line with closures already committed for England, Scotland and Wales. In October 2015, onshore wind developers were offered some respite when DECC

announced that the grace period for RO projects meeting certain criteria would be extended by up to a maximum of nine months, to January 2017.

8.2.2 Feed-in-Tariff

FiTs were introduced in April 2010 with the specific aim to support smaller scale low carbon energy generation and are globally amongst the most popular 'off-the-shelf' policy instruments for incentivizing renewables generation. Generators receive an inflation-linked generation tariff based on generation output for 20–25 years, that is, set at or prior to a project going live. Where electricity is not used locally ('direct supply') but rather exported to the grid, generation tariffs are supplemented with a fixed minimum export tariff or a price of electricity settled through a Power Purchase Agreement (PPA). With their relative predictability and simplicity, FiTs go a long way to removing overall project risk, improving the ease of forecasting cash flow and lowering barriers to entry for non-specialists such as households, community groups and rural businesses. While minimum export tariffs are fixed and technology-blind, FiT generation tariffs are cost and technology-dependent and, together, they have allowed wide and rapid uptake for projects with capacities under 5 MW.

As a technology becomes mainstream and its technology costs decrease, FiT generation tariffs are degressed for new installations. Thus, FiTs have been subject to amendment based on rates of technology uptake. The 2015 reforms differed in that they represented an increase in the speed and extent of degression as well as the possibility of complete removal of FiTs should spending surpass the proposed cap (DECC 2015b). Specifically, the consultation document read: '... *if [proposed changes] cannot put the* [FiT] *scheme on an affordable and sustainable footing then there should be an end to generation tariffs for new applicants as soon as legislatively possible, which we would expect to be January 2016*'. The largest proposed cuts fell on generation tariffs for solar PV and onshore wind, with cuts of up to 87% on prevailing rates (see Table 8.1).

Early parliamentary motions had announced the removal of all subsidies for new onshore wind, but had included explicit statements on the need to protect non-specialist renewable energy generators from reforms. Specifically, Amber Rudd stated: '*I do not wish to stand in the way of local*

Table 8.1 Proposed reductions in feed-in-tariffs, October 2015

Technology	*Band*	*Before*[a]	*After*[b]	*Reduction*
	(kW)	(p/kWh)	(p/kWh)	(%)
Hydro (run-of-river)	0–15	15.45	10.66	31
	15–100	14.43	10.66	26
	100–500	11.40	9.78	14
	500–2000	8.91	6.56	26
	2000–5000	2.43	2.18	10
Solar PV	0–4	12.47	1.63	87
	4–10	11.30	1.63	86
	10–50	11.30	3.69	67
	50–150	9.63	2.64	73
	150–250	9.21	2.64	71
	250–1000	5.94	2.28	62
	1000–5000	5.94	1.03	83
	Stand alone	4.28	1.03	76
Wind (onshore)	0–50	13.73	8.61	37
	50–100	13.73	4.52	67
	100–500	5.89	4.52	23
	500–1500	5.89	4.52	23
	1500–5000	2.49	0	100

[a] *Before* Ofgem tariffs for installations with an eligibility date on or after 1 October 2015;
[b] *After* Proposed Generation Traiffs for January 2016

communities coming together to generate low carbon electricity in a manner that is acceptable to them, including through small scale wind capacity' (Rudd 2015a). The fact that small-scale local projects, including even those adopting the controversial technology of wind turbines, should be singled out raised hopes amongst small-scale renewables developers as well community and local energy practitioners that FiTs might somehow find respite within the storm of reform. Just over a month later, however, DECC published a consultation for the elimination of FiT pre-accreditation—one of the few renewable energy policy provisions that gives leeway specifically to communities (DECC 2015c). The consultation document of 21 July 2015 proposed the removal of pre-accreditation for new anaerobic digestion and hydro of all capacities, and solar PV and onshore wind projects with capacities over 50 kW. Despite a lobbying and petitioning campaign by the fledgling community energy sector, DECC announced on 9 September that it was to remove pre-accreditation by October 2015 (DECC 2015c). The 'pre-registration' option, introduced to protect small school and community (<50 kW) rooftop solar PV projects from tariff degressions, was also removed.

The outcome and deadline for the FiT tariff review were not known at the time of writing, but related policy reforms suggest that the 2015-elected government was broadly not supportive of decentralised and local renewable energy development.

8.2.3 Contract for Difference

Of the triad of major renewable electricity support mechanisms in use in the UK, the CfD is without question the most complex. A CfD is a financial instrument that takes the form of a contract in which a buyer pays a seller the difference between the current value of an asset and its value at contract time and can be thought of as an auction-based Feed-in-Tariff. The application of CfDs to energy markets is unique to the UK and Germany. Theoretically, the value of the asset is determined through a competitive bidding process, and 'the buyer' is the purpose-built government controlled entity called the 'Low Carbon Contract Company' that acts as a subsidiary to DECC. In practice, allocation across several pots for technology clusters of different levels of maturity is very much a negotiated process and has left little room for competitive deployment of lowest cost technologies.

Results of the first allocation round were published on 26 February by DECC and National Grid. Preliminary assessments of expenditures under the LCF suggest that low electricity prices resulted in high payouts under the first auction round of the Contracts for Difference in 2014. Despite this and some other teething problems (related for instance to entry criteria and risk of strategic bids by participants who had hoped to partake under the ROC), the process was widely considered to have been more or less successful. Strike prices were in the realm of £80/MWh for onshore wind, waste-CHP and most solar PV bids (DECC 2015d), and developers started preparing for a second allocation round scheduled for October 2015.

Published CfD allocations for 2015–2023 suggested that relatively small annual budgets would be allocated to mature (CfD "Pot 1") technologies in an attempt to force them to compete.[18] This left the largest proportion of annual CfD allocations for less mature ("Pot 2") technologies, including offshore wind, wave, tidal stream, advanced conversion, anaerobic digestion, dedicated biomass with CHP and geothermal (DECC 2014b). However, preparations for the second CfD FiT round

Table 8.2 Overview of onshore renewable electricity policy mechanism announcements, October 2015

	Technology			
Policy shock	Anaerobic Digestion	Hydro (run- of-river)	Solar PV	Wind (onshore)
ROC termination date			Closure brought forward	Closure brought forward
CfD auction no. 2	Postponed	Postponed	Postponed	Postponed
FiT pre- accreditation	Removed	Removed	Removed	Removed
FiT levels	Consultation expected late 2015	Curtailed, and may be removed, as of January 2016	Curtailed, and may be removed, as of January 2016	Curtailed, and may be removed, as of January 2016

were thrown into disarray following Amber Rudd's announcement on 18 July, which was closely followed by an email from DECC to developers that read: '*There will be no CfD round this October. In the autumn, the Government will set out its plans in respect of the next CfD allocation round*' (Rudd 2015a). The announcement that a second CfD auction would be postponed came in parallel with the previously described cessation of the RO scheme for onshore wind being brought forward. The combined changes raised the spectre of expensive multi-year projects awaiting a planning determination being stranded in a 'support mechanism no-man's land' (see Table 8.2). An emergency meeting hastily convened in Glasgow on 9 July by Scottish Energy Minister Fergus Ewing drew over two hundred wind industry representatives.

8.2.4 Summary of Recent Policy Announcements

In summary, support for dominant technologies (hydro, solar and wind) has been removed altogether for new projects at large scale (ROCs and CfDs) and curtailed for new projects in smaller capacity bands (FiTs). With ROCs for less common technologies set to be phased out across the board by April 2017, the uncertainty surrounding a second CfD auction leaves other large-scale renewables initiatives, including those looking to harness less mature technologies such as wave and tidal, precariously exposed. Announcements about FiT adjustments for anaerobic digestion are expected in late 2015, and here it remains unclear whether changes for smaller scale projects will stretch to other technologies that have thus far seen lower cumulative uptake, such as CHP and geothermal. References in policy documents released to date make it highly unlikely, however, that these technologies will escape unscathed.

Meanwhile, the RHI, which acts much like FiTs do but for heat rather than electricity, has thus far remained conspicuously untouched. However, the RHI was set up to last until March 2016 and is currently under review. With funding for parallel schemes such as the Green Deal Home Improvement Fund removed in July 2015, it is unlikely to be extended. In contrast to the renewable electricity support mechanisms discussed in this chapter, the RHI is not supported through the LCF (see below) but funded directly by the state. As overspending under the latter budget forms the overarching justification for the policy changes to date, the RHI represents something of a litmus test: just how comprehensive will the roll-back on renewables technology support be?

8.3 Explaining 2015 Policy Reforms: Rationale and Historical Context

The policy reforms described here can be characterised as knee-jerk efforts to quell unexpectedly rapid deployment, particularly for onshore wind and solar PV under the RO and FiT schemes, in the context of a limited budget and in which increasing the budget is politically undesirable. The main justification given for FiT and RO reforms was overspending under the LCF and its implications for consumer electricity bills and European Commission State aid approval (Rudd 2015b). In a hearing with the Commons Energy and Climate Change committee, the then new Energy Secretary Amber Rudd stated that '*grid parity projections for the early 2020s were overly pessimistic*' and that '*we should have confidence that renewables will continue to deploy in a way that costs will continue to come down* [without subsidies]', drawing on private discussions with three large onshore wind developers who had stated their willingness to continue building in the absence of subsidy support (Rudd 2015c). Pervasive in the statements and speeches that followed was the notion that the climate debate can and needs to be reclaimed by right-wing politics as an issue that can feasibly be tackled using market-based approaches that are first and foremost 'pro-growth and pro-business' (Rudd 2015a).

While the 2015 policy reforms sent shock waves throughout the renewable energy industry, those who witnessed policy developments leading up the 1990 Electricity Act may recognise the broad lines of an old debate. Historical energy policy reviews, such as those by Mitchell and Connor (2004) and Gross and Heptonstall (2014), characterise UK renewable energy policy as reactive and opportunistic, stemming from a fundamental and persistent lack of consensus within Whitehall over the rationale, objectives and benefits of renewable energy (Gross and Heptonstall 2014; Mitchell and Connor 2004). Although the 2008 Climate Change Act is a manifestation of a long-term cross-party mandate to reduce the national greenhouse gas emissions balance, there was never an equivalent commitment towards renewable energy *per se*. At the core of this long-standing debate is and has always been the question: is there a compelling case to support renewable energy development because it provides benefits that nuclear and gas technologies cannot provide, or are its benefits limited to carbon reduction? This chapter illustrates how this question is at the heart of cross-party issue framing on climate change as being a problem that can be solved through simply internalising a politically acceptable carbon pricing mechanism into a pre-existing institutional framework, or an issue that requires far-reaching institutional reform as well as citizen engagement.

The politics of public intervention around renewable energy extends well beyond UK national boundaries. The reforms would not have been possible were it not for UK's role in eliminating European national renewable energy targets in 2014 on the basis of the need for '*national flexibility to develop a diversified, secure and sustainable energy mix cost-effectively*'. In David Cameron's leaked non-paper for the upcoming 2014 EU Energy Strategy meeting, he argued that climate mitigation and energy security needs should be met through a combination of renewable, CCS, indigenous shale and nuclear projects. The 2015 policy reforms around renewable energy were therefore in part a manifestation of David Cameron's desire to demarcate legislative independence from the European Commission. In this chapter, however, we focus primarily on the key components of the domestic institutional framework governing the UK's energy transition: a revenue-neutral renewable energy subsidy mechanism that finances renewable energy subsidies from a tax on consumer energy bills (the LCF), run in parallel to an imperfect downstream carbon tax on non-domestic consumption for gas, electricity, coal and liquefied petroleum gas (the Climate Change Levy, CCL).

Closer inspection of statements regarding the 2015 energy policy reforms suggests that arguments used to justify reforms around baseload power, affordability, energy security and democratic planning processes were used inconsistently with respect to renewable energy vis-à-vis alternative low carbon technologies (nuclear and shale gas). For instance, subsidy changes were explicitly linked to the commitment to '*give local communities the final say over any new wind farms*' (Rudd 2015d), at the same time as provisions to override local planning authorities on shale gas planning applications were put in place (DECC 2015e). While insulation from external supply shocks is the most prominent aspect of energy security (Watson and Scott 2009) and nuclear and shale gas developments were justified on the basis of energy security (Rudd 2015c), that same logic was not put forward to protect the existing wind or solar industry. Nuclear and shale gas developments were also justified on the basis of providing baseload power (Rudd 2015c), yet mature renewable generation technologies capable of providing baseload power such as biomass and run-of-river hydro (Matek and Gawel 2015) were not exempted from policy reforms. Finally, high levels of financial support for nuclear development at Hinkley Point C were justified on the basis of the immaturity of nuclear technology (Rudd 2015c), while little consideration of the learning curves for renewable energy technologies was in evidence. For example, fledgling technologies such as high-enthalpy geothermal, wave or tidal electricity generation were not singled out as also deserving of special consideration. All in all, these inconsistencies suggest that objectives around energy security, grid stability and local preferences may not in fact be driving these energy policy reforms, as much as being framed to fit an incoherent and multi-stranded anti-renewables ideology. Furthermore, since August 2015 renewable generators have been made subject to the CCL, raising serious doubts over the current government's commitment to climate mitigation more generally.

The arguments put forward to justify the reforms highlight how public discourse around renewable energy subsidies is dominated not by the relative benefits of different energy technologies but by their relative cost, the implications for heat and electricity prices, their visual intrusiveness and repercussions on the competitiveness of UK industry. The rationale is strongly aligned to neoclassical regulation theory in that it is seemingly singularly geared towards minimising the per-kWh cost of delivering low carbon energy, and sidesteps considerations to do with technological lock-in, the nurturing of a domestic renewable energy industry, or the potential of distributed energy to generate local benefits that are not accounted for through energy bills. Even here, however, the reforms are not consistent with cost-efficiency arguments insofar as nuclear technologies are capital-intensive and not unambiguously more competitive than solar or wind, as demonstrated by the outcome for the first round of the CfD FiT allocations (DECC 2013a). Furthermore, significant uncertainties surround both their final delivery timescale and capital cost, as demonstrated by severe problems in meeting construction timescales and budgets for identical EPR reactors at Flamanville and Olkiluoto. High levels of financial support for nuclear development were justified on the basis that baseload power requires a price premium (Rudd 2015c), but baseload power plants produce electricity at lowest cost.

A report by Howard (2015) entitled 'The Customer is Always Right: Putting consumers back at the heart of UK energy policy' endorsed by the Energy Secretary provides recommendations that are somewhat consistent with the 2015 policy reforms and sheds light on what appears to be the underlying rationale. Howard suggests that the low carbon energy objective has trumped the objective to provide consumers with affordable energy. Specifically, he argues that network costs and clean energy policy costs managed by DECC under the LCF are responsible for 50% of price hikes in both gas and electricity during the period 2009–2014, which increased by 185 and 120%, respectively. Based on the notion that renewable energy subsidies have been implemented with a lack of centralised overview and consideration for consumer energy bills, he goes on to recommend the halt of public support for '*expensive technologies [including] wave and tidal stream, solar thermal and heat pumps*', suggesting that government ought to '*pick winners over propping up losers*' (Howard 2015). His recommendation is for competitive and technology-neutral allocation of subsidies under the CfD to support low-cost mature technologies such as medium and large-scale solar PV and wind, with limited support for pre-commercial technologies and/or small-scale installations, pending their ability to demonstrate potential for cost reduction. While policy statements fully embraced Howard's line of reasoning (DECC 2015c; Rudd 2015b), the reforms that were implemented have done the opposite. Specifically, they have pushed lowest cost renewable energy technologies such as onshore wind out of the market altogether, continued to provide support to small-scale less mature technologies, and put in place ambiguity over the continuation of support for less mature technologies above 5 MW. It is a moot point as to whether Howard's recommendations find expression in the high levels of public support given to immature technologies and nuclear power developments at Hinkley Point.

The emerging rationale is characteristic of technology-neutral approaches to climate mitigation policy, in that it does not take account of the systemic factors that constrain and enable the emergence of cost-competitive technology—including requirements and compatibilities of different technologies in relation to demand profiles, structure of the existing power market supply, risk premiums facing new technologies or externalities that are a function of increased adoption (such as innovation and learning spillovers, imperfect competition, supply chain co-ordination effects, and legitimacy costs) (Foxon 2005; Kalkuhl et al. 2013; Lehman et al. 2012). These factors drive internal and external increasing returns-to-scale that underlie technology learning curves at various stages of maturity. The role of technology learning curves in determining costs has meant that the cost-efficiency of support policies has been higher where they have effectively facilitated rapid deployment and industry consolidation (International Renewable Energy Association (IRENA), 2015; Lauber 2015). The decision of a national government not to stimulate a domestic renewable energy market therefore reflects a lack of confidence in the ability of its domestic industry to compete on the national and international stages, representing instead a wait-and-see policy that depends on technology and service imports (Gross and Heptonstall 2014). For example, technology learning curves for solar PV continue to surpass predictions, exhibiting 75% cost reductions since 2009 (IRENA 2015). Following pure cost-efficiency arguments would have required dismissing early stage deployment of solar PV and foregoing the development of a domestic market. In summary, while cost-efficiency and affordability are legitimate concerns, taken alone they are not a sufficient basis on which to evaluate public intervention for pre-commercial technologies. A more constructive policy assessment would focus on the design of the feedback and decision-making mechanisms that enable DECC to provide transparency and long-term stability around demand incentives, while enabling it to monitor learning curves and allocate resources to highest potential technologies. Such a mechanism would acknowledge the unique benefit and constraints of contending technologies to provide low carbon heat and electricity and enable informed resource allocation that takes into account estimated short- and long-term costs and benefits.

The singular focus on cost-efficiency driving the 2015 policy reforms does not bode well for future support for small-scale technologies under the FiT and RHI schemes, for which public support is often legitimised on the basis of a number of soft positive externalities and indirect effects. Due to the distributed nature of renewable energy generation, renewable energy development is able to attract capital locally and generate socio-economic benefits over and above those brought by nuclear and gas technologies that merit public support. For instance, there is evidence that it has spurred new domestic industry in rural areas with longer-term knock-on effects, resulting in income diversification and rural socio-economic resilience (Allan et al. 2011). Engaging the public in energy generation is thought to generate awareness and buy-in for renewable energy (Hvelplund 2006; Walker and Cass 2007), a notion that is supported by the fact that democracies with high levels of renewable energy deployment such as Germany, Denmark and Sweden also demonstrate high levels of local ownership (Roberts and Bodman 2014). Furthermore, historic breakthroughs in cost-reduction have come from FiT-supported small-scale renewable energy technologies (Fouquet and Johansson 2008; Lauber 2015). This suggests that public support for pre-commercial small-scale technologies pays off in the form of technological learning, price reductions, new employment and exports. In conclusion, if the rationale is to achieve socially optimal public interventions in the energy sector, portraying renewable energy subsidies under the LCF as a public over-investment on a £-per-kWh basis rests on a short term and incomplete analysis, but above all on a decision to exclude or discount benefits that are more challenging to internalise in the market.

The analysis so far suggests that within this current institutional framework, the UK has limited its rate of renewable electricity deployment to the rate at which subsidies can perceivably be levied directly from energy consumers. However, there has been no inclusive and transparent dialogue around public acceptance of renewable energy levies, nor on the larger question of who should fund large infrastructural transitions from a social justice standpoint (Dresner et al. 2006). A great deal is known about the relative theoretical efficiency and welfare effects of different mitigation policies, including distributional and energy price effects (Fisher 2010; Kalkuhl et al. 2013; Pearce 2006). Because renewable energy penetration reduces average wholesale electricity costs, temporary renewable energy subsidies in fact serve to ease distributional effects of taxes on carbon at reasonable additional cost and are seen by many economists as a second-best alternative to perfect (but unpopular) carbon taxes. Where renewable energy subsidies are financed from non-renewable energy production taxes, electricity prices will not rise as long as the renewable energy sector is able to respond to electricity price changes and displace non-renewable generators (Pearce 2006). This means that the efficiency and effectiveness of renewable energy subsidies cannot be assessed without also examining the UK CCL.

Non-domestic high energy consumers are eligible for 80% reduction on the CCL, and the CCL is currently recycled into the industry in the form of savings on labour insurance. With fossil fuel generators subject to substantial tax increases under the Carbon Price Support since April 2015 (HM Revenue and Customs 2015) and renewable generators newly subjected to the CCL (since August 2015), the CCL appears to be moving ever closer to a general energy tax. Even where effective carbon prices are in place, however, consumers can only respond to carbon price signals if technology-specific policies succeed in making low carbon alternatives available (Anderson et al. 2001). Furthermore, a recent cross-country comparative study has shown that demand pull and supply push policies in low carbon industries that successfully engage the wider public in renewable energy generation are likely to bring voters and economic constituencies into coalitions for decarbonisation, which subsequently fosters political support for more comprehensive carbon pricing policies (Meckling et al. 2015). In summary, if the current government was committed to climate mitigation, it would need to revisit its ambitions for a carbon tax and consider alternative financing options in which renewable production subsidies and associated transmission and balancing costs are distributed across a broader range of actors.

While it is easy to find fault in government policy, coaxing the power market through a low carbon transition is a difficult balancing act. Where generation incentives effectively attract renewable energy deployment, average wholesale electricity prices decrease and further renewable energy penetration risks pushing existing balancing services with higher marginal cost out of the electricity market (Klessmann et al. 2008). The National Grid is anticipating continued increase in embedded generation (up from 8 to 16 GW from 2014 to 2016), expressing concerns over its ability to put in place coping mechanisms to prevent grid management issues around forecasting, inertia and frequency response (National Grid 2015). Given low gas and oil prices and declining wholesale electricity prices since October 2014, the UK government is likely under pressure to protect industries supplying balancing power in addition to reining in upward pressure on electricity retail prices. Nevertheless, the 2015 policy reforms demonstrate the fragility and temporality of policy paradigms and public support frameworks enabling renewable energy deployment. Clearly, previous studies outlining a 'paradigm shift' in UK energy policy underestimate the role of deep-seated cross-party politics and the mutability of institutional reform (Holburn 2012; Kern et al. 2014).

In summary, it was the scrapping of national renewable energy targets under the European Renewable Energy Directive combined with the recession and the 2015 Conservative Party election outcome that created the political opportunity for parliamentary discourse to sway back towards technology-neutral carbon mitigation policies. Despite targets to eliminate coal power by 2023, the speed and extent of reforms singularly targeting renewable energy technologies strongly suggests that consumer protection and short-term cost-efficiency have taken precedence over climate mitigation. It is possible that such wait-and-see policies will become more prominent as conservative resistance to public support frameworks take hold and Europe lowers its climate mitigation ambitions (Wyns et al. 2014). Wait-and-see policies may also be encouraged by the emergence and consolidation of global renewable energy industry leaders, which leaves new entrants too far behind to catch up, especially for renewable energy technologies that are some years off grid parity. In the following section, we summarise the likely implications of the 2015 policy reforms for the Scottish renewable energy industry.

8.4 Conclusion: Exploring the Implications of 2015 Renewable Energy Reforms

Although some elements of renewables regulation such as planning are controlled from Holyrood, the key regulatory powers in terms of design, grace periods and eligibility criteria for renewable energy support incentives are governed by the UK Government as per the 2013 Energy Act. Under the RO, the Scottish Government used its power over grandfathering and support levels to maintain certificate prices and grandfathering policies for solar PV until April 2016, but stated it did not have the power to prevent a Whitehall imposed closure after April 2016 (Ewing 2015), demonstrating how Scottish ambitions with respect to renewables development are largely at the mercy of Whitehall politics.

A large body of literature demonstrates the effects of the premature withdrawal of policy support schemes and financial subsidies on renewables deployment (Barradale 2010; Dél Rio and Tarancon 2012; Meyer and Kocfoed 2003; Nemet 2010). The nature of many renewable energy technologies brings considerable challenges to the design of robust, cost-effective support mechanisms. Renewable energy projects are characterised by large upfront investment costs, where energy yields needed to cover early capital outlays are often at the mercy of imperfectly predictable environmental conditions. Development timescales from inception to operation are generally in the order of years, easily outlasting election cycles and associated changes in dominant political agendas. This generic project risk profile carries a profound implication; from conception to financial close, the risk of renewables support mechanisms being adjusted or withdrawn can make or break a project. This is particularly true for less mature technology markets where markets and supply chains have not yet fully formed, and expertise may still be lacking, such that costs are high and uncertain. The risk of outright 'renewables policy mechanism default', in which policy interventions would affect actual operational projects, is carried throughout the expected lifetime of the project (15–25 years for most technologies). Fundamentally, for the current support mechanisms to be effective, developers and finance providers need to have faith in their continuity (Lüthi and Prässler 2011; Ulph and Ulph 2013).

While it is too early to assess the impacts of these changes on the Scottish (and wider UK) renewables sector empirically, it is possible to make an informed guess as to the impacts of these reforms, drawing on past policy shocks in Spain and Australia to provide some clues as to the fate of renewable energy development upon rapid removal of public support schemes. Deployment will decline after a rush to complete projects already in the pipeline within existing windows of opportunity which are determined by degressing FiTs, loss of pre-accreditation as well as the nine-month grace period for project eligibility for the RO after April 2016. Pending a second CfD auction round, the push towards a decarbonised energy sector may largely come to a halt once existing projects are completed. In a more likely scenario, CfD auctions are continued (Rudd 2015c). However, annual CfD allocations are relatively small, ranging from 10 to 50% of past annual RO expenditures (DECC 2013b), and are likely to exclude onshore wind (Rudd 2015c). In this scenario, overall annual deployment rates will decrease substantially compared to the period 2009–2016. Despite the government's repeated emphasis on 'technology-neutral support', the reforms described in this chapter suggest that the UK government will in future be highly selective in those technologies that it supports. Based on CfD and FiT support in place at the time of writing, piecemeal development in offshore wind, geothermal, tidal and wave initiatives seem likely. Temperton and Schoenberg (2015) suggest that there is simply no LCF budget to ease less mature technologies gradually into the market, and that there is no budget for the majority of the offshore wind projects in the pipeline under the current allocation, such that they are de facto competing over remaining budget with far less mature technologies such as CCS and tidal projects. In this scenario, deployment may not drive learning and cost-efficiency improvements required for widespread diffusion. Mirroring events following policy shocks in Spain and Australia, it is likely that large established companies will look to expand renewable energy markets elsewhere, such as North America, Brazil, Chile and Mexico. Recent policy statements imply additional support for innovation around storage (Temperton and Schoenberg 2015) in an attempt to overcome continued grid capacity constraints associated with further penetration of intermittent renewable energy. Finally, deployment under proposed FiT rates for solar PV and onshore wind is likely to be restricted to remaining high capacity sites, although an anticipated drop in solar PV prices due to the expiry of import tariffs in December 2015 may improve the financial viability of solar PV projects.

Employment will decline as firms adjust and downsize to new market conditions, but some sectors may increase employment temporarily in the rush to complete projects before funding cliff-edges. The renewable energy sector in the UK employs around 112,000 people across the value chain in 2014 (Renewable Energy Association 2015), with over 21,000 direct jobs in generation and project development in Scotland (2015). Many of these jobs are located in the remote parts of Scotland, where renewable energy developments have supported economic diversification of rural economies.

It is hard to see how high levels of Scottish renewables deployment could continue in the absence of cross-border subsidy from UK consumers. A UK energy market and continued reliance on UK energy policies is part of the uncomfortable nature of the devolution settlement. The Smith Commission (2014) sets out proposals for '*a formal consultative role for the Scottish Government and the Scottish Parliament in designing renewables incentives and the strategic priorities set out in the Energy Strategy and Policy Statement to which OFGEM must have due regard*'. The post-Smith Commission White Paper alludes to the importance of not '*sacrificing the integrity or stability of the GB-wide energy market*'. However, there was little or no formal consultation and very little influence of the Scottish Government's renewable policy priorities on the emergent renewables policies at the UK level (Ewing 2015).

It is probable that the survivors amongst the renewables companies will seek innovative ways to sustain activity, for instance through efficiency improvements in supply chains. Where there are significant buyers close to renewable energy production capacity, selling directly to final consumers and undercutting grid-provided electricity prices may offer a solution, in particular where cost-effective storage enables larger capacity installations and local grid solutions. For instance, if councils were to purchase local energy for schools or social housing, or private sector high energy consumers could connect to local suppliers, there may be possibilities of win–win outcomes. Large hotels in rural areas or energy intensive businesses provide nodes of demand that local renewables suppliers could connect to. Similarly, business models based on self-consumption for domestic housing and businesses are likely to increase and capacity size of installations likely to decrease accordingly. However, the loss of the security of income streams is likely to limit the scale of deployment.

Where might the Scottish Government have some influence? It can pump prime initiatives to provide novel solutions to energy storage, which are particularly important in isolated and island communities. It can fund initiatives exploring the deepening of opportunities for local grid provision and it can continue to fund initiatives with respect to community energy, albeit without the support of FiTs. These mostly small-scale action research projects are valuable pilots, but unless the pilots can show evidence of secure returns to investors, their further roll-out is likely to be constrained. The Scottish Government may oppose austerity politics but it lives in its shadow and its funding is constrained by austerity. European funding streams may offer a lifeline for some types of development.

The 2014–2020 Rural Development Programme at European level offers incentives for renewables development. The new Scottish Rural Development Programme (RDP) does not include these because it had almost certainly worked on the assumption that FiTs would trump RDP grant aid, but both farmers and community groups are potentially eligible for funding should measures be included. The European Social Fund and ERDF money could also be targeted at renewables support.

Energy policy is not unconnected to fuel poverty, and the recent policy desire to limit exchequer support for renewables connects to this issue. Fuel poverty has increased in Scotland almost threefold since 2002 to affect just under 40% of households. The Scottish Government launched a new scheme in April 2015 to '*install insulation*, heating and low *carbon or renewable measures in the homes of households who are identified as living in fuel poverty, with a wider range of options for people living off the main gas grid including solar thermal and biomass systems*'. As fuel poverty is especially high in more remote rural areas, connecting renewables development to reducing fuel poverty reduction could be an explicit goal of community-led renewables.

The experiences of the last turbulent year have exposed the vulnerability of the Scottish renewables sector to major UK policy shifts. Looking forward, it is difficult to see any prospect of a devolved Scottish energy market. In its absence, policy diktats from the UK government will continue to frame Scottish possibilities. Some large-scale developments will continue, particularly in the marine-based technologies. And, at the margins, there is likely to be a little wriggle room in which Scotland can pilot novel approaches and keep the sector alive in ways that must of necessity be more cost effective, must deliver benefit to energy users and must sustain energy generators with reasonable returns.

Few anticipated the depth and breadth of the assault on UK renewables policy that will with certainty have devastating consequences on the sector. The early UK government rhetoric around renewables policy reform in 2015 still found space for support for community renewables but even this was subjected to a flood tide of policy change that has left much of the renewables sector with little or no confidence in future. Those in the sector were aware of the need for policy support degression and worked within those constraints. What they were unaware of was just how far a more doctrinaire (if at times inconsistent) neoliberalism had penetrated the new political administration and how much coalition politics had tempered the support for renewable energy in the 2010–2015 coalition government.

References

Allan, G., P. Mcgregor, and L. Swales. 2011. The Importance of Revenue Sharing for the Local Economic Impacts of a Renewable Energy Project: A Social Accounting Matrix Approach. *Regional Studies* 45 (9): 1171–1186.

Anderson, D., C. Clark, T.J. Foxon, R. Gross, and M. Jacobs. 2001. *Innovation and the Environment: Challenges and Policy Options for the UK*. London: Imperial College Centre for Energy Policy and Technology & the Fabian Society.

Barradale, M. 2010. Impact of Public Policy Uncertainty on Renewable Energy Investment: Wind Power and the Production Tax Credit. *Energy Policy* 38 (12): 7698–7709.

DECC. 2013a. Electricity Generation Costs—December 2013 (online). https://www.gov.uk/government/uploads/system/uploads/attachment_data/file/269888/131217_Electricity_Generation_costs_report_December_2013_Final.pdf. Accessed 17 Sept 2015.

DECC. 2013b. EMR Delivery Plan, December 2013 (online). https://www.gov.uk/government/publications/electricity-market-reform-delivery-plan. Accessed 15 Oct 2015.

DECC. 2014a. Government Response to Consultation on Changes to Financial Support for Solar PV—2 Oct 2014.

DECC. 2014b. Draft Budget Notice for CFD Allocation Round 1 (online). https://www.gov.uk/government/publications/indicative-cfd-budget-notice-for-the-autumn-2014-cfd-allocation-round. Accessed 1 Oct 2015.

DECC. 2015a. Consultation on Changes to Feed-in Tariff Accreditation—21 July 2015.

DECC. 2015b. Consultation on a Review of the Feed-in Tariffs Scheme—27 Aug 2015.

DECC. 2015c. Controlling Spending on Solar PV Projects of 5 MW and Below Within the Renewables Obligation—22 July 2015.

DECC. 2015d. Contracts for Difference (CFD) Allocation Round One Outcome (online). https://www.gov.uk/government/uploads/system/uploads/attachment_data/file/407059/Contracts_for_Difference_-_Auction_Results_-_Official_Statistics.pdf. Accessed 7 Sept 2015.

DECC. 2015e. Faster Decision Making on Shale Gas for Economic Growth and Energy Security (online). https://www.gov.uk/government/news/faster-decision-making-on-shale-gas-for-economic-growth-and-energy-security. Accessed 27 Oct 2015.

Dél Rio, P., and M. Tarancon. 2012. Analysing the Determinants of On-Shore Wind Capacity Additions in the EU: An Econometric Study. *Applied Energy* 95: 12–21.

Dresner, S., T. Jackson, and N. Gilbert. 2006. History and Social Responses to Environmental Tax Reform in the United Kingdom. *Energy Policy* 34: 930–939.

EU Commission. 2015. Commission Renewable Energy Progress Report (online). http://ec.europa.eu/energy/en/topics/renewable-energy/progress-reports. Accessed 2 Sept 2015.

Ernst and Young. 2015. Renewable Energy Country Attractiveness Index, Issue 45: September 2014 (online). www.ey.com/Publication. Accessed 2 Sept 2015.

Ewing, F. 2015. Renewables Obligation Scotland—Solar PV Projects of 5 MW and Below, Open Letter from the Scottish Government, 22 September 2015 (online). http://www.gov.scot/Resource/0048/00485675.pdf. Accessed 20 Oct 2015.

Fisher, C. 2010. Renewable Portfolio Standards: When Do They Lower Energy Prices? *The Energy Journal* 31 (1): 101–120.

Fouquet, D., and T. Johansson. 2008. European Renewable Energy Policy at Crossroads: Focus on Electricity Support Mechanisms. *Energy Policy* 36 (11): 4079–4092.

Foxon, T. 2005. UK Innovation Systems for New and Renewable Energy Technologies: Drivers, Barriers and Systems Failures. *Energy Policy* 33 (16): 2123–2137.

Gross, R., and P. Heptonstall. 2014. Time to Stop Experimenting with UK Renewable Energy Policy. ICEPT Working Paper, October 2010.

HM Revenue and Customs. 2015. A Guide to the Carbon Price Floor, (online). https://www.gov.uk/government/publications/excise-notice-ccl16-a-guide-to-carbon-price-floor/excise-notice-ccl16-a-guide-to-carbon-price-floor. Accessed 29 Oct 2015.

Holburn, G. 2012. Assessing and Managing Regulatory Risk in Renewable Energy: Contrasts Between Canada and the United States. *Energy Policy* 45: 654–665.

Howard, R. 2015. The Customer is Always Right: Putting consumers back at the heart of UK energy policy—Policy Exchange Report (online). https://policyexchange.org.uk/?s=the+customer+is+always+right. Accessed 10 Oct 2015.

Hvelplund, F. 2006. Renewable Energy and the Need for Local Energy Markets. *Energy* 31: 1957–1966.

IRENA. 2015. Renewable Power Generation Costs In 2014—January 2015 (online). https://www.irena.org/DocumentDownloads/Publications/IRENA_RE_Power_Costs_2014_report.pdf. Accessed 10 Oct 2015.

Kalkuhl, M., O. Edenhofer, and K. Lessmann. 2013. Renewable Energy Subsidies: Second-Best Policy or Fatal Aberration for Mitigation? *Resource and Energy Economics* 35: 217–234.

Kern, F., C. Kuzemko, and C. Mitchell. 2014. Measuring and Explaining Policy Paradigm Change: The Case of UK Energy Policy. *Policy & Politics* 42 (4): 513–530.

Klessmann, C., C. Nabe, and K. Burges. 2008. Pros and Cons of Exposing Renewables to Electricity Market Risks-A Comparison of the Market Integration Approaches in Germany, Spain, and the UK. *Energy Policy* 36 (10): 3646–3661.

Lauber, V. 2015. Political Economy of Renewable Energy. *International Encyclopaedia of the Social & Behavioural Sciences* 18: 367–373.

Lauber, V., and S. Jacobsson. 2016. The politics and economics of constructing, contesting and restricting socio-political space for renewables – the German Renewable Energy Act. *Environmental Innovation and Societal Transitions* 18: 147–163.

Lehman, P., F. Creutzig, M.H. Ehlers, N. Friedrichsen, C. Heuson, L. Hirth, and R. Pietzcker. 2012. Carbon Lock-Out: Advancing Renewable Energy Policy in Europe. *Energies* 5: 323–354.

Lüthi, S., and T. Prässler. 2011. Analyzing Policy Support Instruments and Regulatory Risk Factors for Wind Energy Deployment—A Developers' Perspective. *Energy Policy* 39: 4876–4892.

Mackenzie, W. 2009. Scotland's Generation Advantage—A report to the Scottish Government (online). http://www.scotland.gov.uk/Resource/Doc/295424/0091448.pdf. Accessed 25 Aug 2011.

Matek, B., and K. Gawel. 2015. The Benefits of Baseload Renewables: A Misunderstood Energy Technology. *The Electricity Journal* 28 (2): 101–112.

Meckling, J., N. Kelsey, E. Biber, and J. Zysman. 2015. Winning Coalitions for Climate Policy. *Science* 349: 1170–1171.

Meyer, N., and A. Koefoed. 2003. Danish Energy Reform: Policy Implications for Renewables. *Energy Policy* 31 (7): 597–607.

Mitchell, C., and P. Connor. 2004. Renewable Energy Policy in the UK 1990–2003. *Energy Policy* 32: 1935–1947.

National Grid. 2015. Electricity Customer Seminar, Glasgow, 6 Oct 2015.

Nemet, G. 2010. Robust Incentives and the Design of a Climate Change Governance Regime. *Energy Policy* 38: 7216–7225.

OFGEM. 2011. Renewables Obligation: Guidance for Licensed Electricity Suppliers (GB and NI), (online). https://www.ofgem.gov.uk/publications-and-updates/renewables-obligation-guidance-licensed-electricity-suppliers-may-2013. Accessed 25 Oct 2013.

Pearce, D. 2006. The Political Economy of an Energy Tax: The United Kingdom's Climate Change Levy. *Energy Economics* 28: 149–158.

Renewable Energy Association. 2015. Annual Report on the UK Renewable Energy Sector, (online). http://www.r-e-a.net/resources/rea-publications. Accessed 22 Sep 2015.

Roberts, J., and F.R. Bodman. 2014. Community Power-Model Legal Frameworks for Citizen-Owned Renewable Energy. *ClientEarth*. https://www.clientearth.org/reports/community-power-report-250614.pdf. Accessed 31 May 2017.

Rudd, A. 2015a. *Ending New Subsidies for Wind*, 18 June 2015. Written statement to Parliament.

Rudd, A. 2015b. Speech on Climate Change—24 July 2015. In *Aviva Conference*, London.

Rudd, A. 2015c. Commons Select Committee on Energy and Climate Change Questions he Secretary of State as Part of the Committee's Inquiry into the Department of Energy and Climate Change's (DECC) priorities for 2015 (online). http://www.parliament.uk/business/committees/committees-a-z/commons-select/energy-and-climate-change-committee/news-parliament-2015/decc-priorities-2015/. Accessed 22 Sept 2015.

Rudd, A. 2015d. Statement from the Secretary of State for Communities and Local Government, 18 June 2015 (online). http://www.parliament.uk/documents/commons-vote-office/June%202015/18%20June/1-DCLG-Planning.pdf. Accessed 22 Sept 2015.

Temperton, I., and M. Schoenberg. 2015. The Levy Control Framework Beer Mat, March 2015 Climate Change Capital Report (online). http://www.climatechangecapital.com/thinktank/research/your-levy-control-framework-beer-mat. Accessed 1 Mar 2015.

The Smith Commission. 2014. Report of the Smith Commission for Further Devolution of Powers to the Scottish Parliament—27 November 2014 (online). http://www.smith-commission.scot/wp-content/uploads/2014/11/The_Smith_Commission_Report-1.pdf.

Ulph, A., and D. Ulph. 2013. Optimal Climate Change Policies When Governments Cannot Commit. *Environmental & Resource Economics* 56 (2): 161–178.

Walker, G., and N. Cass. 2007. Carbon Reduction, 'The Public' and Renewable Energy: Engaging with Socio-Technical Configurations. *Area* 39: 458–469.

Watson, J., and A. Scott. 2009. New nuclear Power in the UK: A Strategy for Energy Security? *Energy Policy* 37: 5094–5104.

Wyns, T., A. Khatchadourian, and S. Oberthür. 2014. EU Governance of Renewable Energy post-2020—Risks and Options, A Report for the Heinrich-Böll-Stiftung European Union (online). www.ies.be/files/eu_renewable_energy_governance_post_2020.pdf. Accessed 25 Sept 2015.

Authors' Biography

Anna L. Berka is a graduate of the Universities of Oxford and Wageningen, with M.Sc. degrees in Economics and Environmental Science. She is an Early Stage Researcher with the European Union Marie Curie Innovative Training Network, Ph.D. candidate with the Department of Forest Economics at the University of Helsinki, as well as co-founder and research associate with Scene Consulting. She has led numerous events under the banner of social and organisational learning for the greening of both public and private enterprises, and coordinated the UK's first community energy database initiative in 2011–2012.

Jelte Harnmeijer A systems thinker and practitioner who strongly believes in integrating research with implementation, Jelte specialises in distributed low-carbon energy solutions. Jelte is the T.B. Macaulay Renewables Fellow at the James Hutton Institute, where he focuses on how community renewables can help address pressing challenges such as food and fuel poverty, inequality, and global climate destabilisation. He is a Founding Partner at Scene Consulting, an Edinburgh-based social enterprise that assists non-specialists in meeting their renewable energy development objectives. He holds concurrent fellowships in Carbon Economics and Carbon Finance at the University of Edinburgh.

Bill Slee is an Emeritus Fellow at the James Hutton Institute. His main interests include evaluation of rural development policies, including agri-environment policy, climate change and land use, forestry and community energy. He is active on two EU Horizon 2020 projects on innovative delivery of public goods and social innovation.

CHAPTER 9

Trouble on the Horizon? Further Devolution and Renewable Electricity Policy in Scotland

Geoffrey Wood

9.1 Introduction

With further devolution of powers firmly on the table, a window of opportunity to discuss a new devolutionary settlement exploded onto the political and public arena, offering the prospect of substantial changes for energy in terms of both policy and practice in Scotland. Energy has long been a highly contentious political issue since the discovery of commercially viable hydrocarbon reserves in the Scottish North Sea (Harvie 1995). With increasing concerns over climate change, security of supply and potential economic gains from developing domestic and export capabilities and resultant employment, the renewable electricity sector is a key focus in any subsequent debate.

Scotland has evidenced significant growth in renewable electricity technology deployment. Since winning the 2007 election, the SNP-led Scottish Government has focused heavily on renewable energy as an integral part of its sustainable economic strategy and as a core argument in the SNP's move towards independence from the UK (Scottish Government 2008). In the end, the Scottish people voted to remain

G. Wood (✉)
School of Law, University of Stirling, Stirling, UK
e-mail: geoffrey.wood@stir.ac.uk

G. Wood and K. Baker (eds.), *A Critical Review of Scottish Renewable and Low Carbon Energy Policy*, Energy, Climate and the Environment,
DOI 10.1007/978-3-319-56898-0_9

within the UK. With the vote significantly closer than many believed possible on both sides of the independence debate (55–45% in favour of remaining), commitments made in the final stage of the referendum by the main pro-unionist parties, Labour, the Liberal Democrats and the Conservatives reignited calls for increased powers and the development of a new substantially enlarged devolutionary settlement.

As discussed by Wood elsewhere in this book, there are a number of concerns with the existing devolutionary settlement regarding the RET sector and indeed wider energy issues: it is largely individualistic, piecemeal and arbitrary in terms of what is reserved and devolved to the UK and Scottish Governments, respectively. Chapter 2 (Wood) shows that the Scottish Government currently has neither a comprehensive nor a cohesive set of devolved powers over renewable electricity policy and practice. With the need to deploy around 8 GW of new capacity over the next 5 years from around 8 GW of capacity today (2016), requiring a step change in the deployment of offshore wind critical to the success of meeting the demanding 2020 target of 100% equivalent of gross electricity consumption from renewable energy sources, this has created a mixed bag of opportunities and challenges to RET deployment in Scotland and the wider UK. In an area, where policy, legislative and regulatory activity 'has accelerated almost breathlessly' (Pearson and Watson 2010: 3), there is an added urgency in understanding the implications for renewable electricity. Under a strict timetable, the Smith Commission published its recommendations for further devolution in the Report of the Smith Commission for further devolution of powers to the Scottish Parliament (hereafter the Smith Report) (Smith Commission 2014), with the resultant Scotland Act 2016 passed by Parliament in March 2016. Importantly, the debate appears by no means settled, with the options falling significantly short of the Scottish Government's aspirations as set out in the post-referendum document More Powers for the Scottish Parliament: Scottish Government Proposals (Scottish Government 2014a, b).

This chapter assesses the relevant provisions of the Scotland Act alongside both the Scottish Government's aspirations pre- and post-referendum and the implications for renewable electricity technology deployment with regard to the 2020 target and beyond. In particular, this chapter aims to answer whether or not the latest Scotland Act will result in a more cohesive set of devolved powers to the Scottish Government in the key area of renewable electricity.

9.2 A Vision for a Yes Vote: The White Paper on Independence

In the event of a yes vote, the Scottish Government set out its post-referendum independence vision for the shape of the energy sector in the 2013 document Scotland's Future—Your Guide to an Independent Scotland (hereafter the White Paper on Scottish Independence). Although the campaign for independence was lost, it is worthwhile examining the proposals, not merely as an exercise in *what if*, but in order to set in context the proposals for further devolution; both independence and further devolution call for increased responsibility to the Scottish Government, albeit to different degrees, and the White Paper on Independence highlighted the need for consistency and change in any subsequent transfer of powers given the high level of connectedness in the energy system between Scotland and the rest of the UK.

Independence raises important questions of ownership of both the assets and obligations. Energy infrastructure, including generating plant and transmission and distribution wires, is expensive long-lived assets lasting between 25 and 50 years or more. A lot of the UK energy infrastructure is already in place and operational. Security of supply concerns, climate change and renewable energy objectives and the requirement to close coal-fired power stations for environmental reasons and an ageing nuclear fleet in the near future mean that more capacity, especially low carbon energy and renewables, is needed to be brought online to fill the gap and meet legally binding obligations. One of the key issues during the referendum centred on who 'owns' the existing and future assets and the targets and obligations: How would this be divided up between Scotland and the rest of the UK (rUK)? Would ownership of the costs, liabilities and responsibilities be shared across an independent Scotland and the rUK? If the costs are not socialised, what would the impact be on Scottish consumers? These were important issues during the referendum that continue to resonate throughout the ongoing debate on further devolution. Furthermore, it needs to be recognised that a number of these responsibilities are of an international nature, for example, the EU climate change and renewable energy directives. In addition to the security of supply and climate change concerns, responsibilities also include consumer bills and the issues of affordability and fuel poverty, both highly emotive and political topics.

The independence referendum campaign also revealed the contradictory and somewhat paradoxical nature of the relationship between the UK and the Scottish Government. Despite making it clear that it wanted Scotland to remain within the union, in stating the UK position on independence (or rather, no independence) and energy, the UK Government publication *Scotland Analysis: Energy* is a curious document. On the one hand, it states the important contributions made by Scotland: that the fully integrated GB-wide energy market is very successful and beneficial for the UK, that over 40% of total private sector investment in the last four years has occurred or is planned to occur in Scotland (an estimated £14 billion out of a UK total of £34 billion) and that Scotland punches well above its weight in terms of the proportion of RET deployment and generation output in Scotland: 34% of all the UK's RET capacity and 32% of all RES-E generation in 2013. On the other hand, the document goes on to state that this is problematic: '*Spending via the Renewables Obligation reached £2 billion in 2012–2013. Of this Scotland received £560 million, which is 28% of the total* [RO] *funding. This is a considerable amount given Scotland only accounts for around 10% of current electricity sales in the UK*' (HM Government 2014: 52). This is a case of comparing apples with oranges; the sale of electricity and the level of subsidy offered on a country-specific basis have nothing to do with the generation of RES-E. Although the UK document justifies its stance by stating that as independent states, the focus for Scotland and the rUK would be on country-specific policy objectives and that there would be no requirement on the rUK for Scottish-sourced RES-E generation, an arguably unsubstantiated but nevertheless important point, there is no discussion of potential areas of cooperation or how the rUK would address concerns, namely security of supply and meeting the 2020 renewable energy target.

One of the key proposals of the White Paper on Independence is that Scotland would continue to participate in the existing GB-wide market for electricity and gas under the existing single Transmission Operator (National Grid). In return for Scotland's significant renewable energy resources, existing generation capacity and the transfer of electricity to the rUK, the Scottish Government would seek '*... a far greater degree of oversight of the market arrangements for energy and firmer safeguards over Scottish energy security*' (Scottish Government 2013: 295). The overall approach to energy would be directed by the establishment of a jointly controlled Energy Partnership with Westminster, with a new independent national regulatory body taking over the relevant functions for Scotland but working closely with the pan-rUK regulator OFGEM. Importantly, in return for Scottish renewable energy and security of supply benefits, the UK Government would continue the system of shared financial support for renewables and capital costs for grid infrastructure across the approximately 26 million households in comparison to less than three million in Scotland.

The question becomes very much one about the level of influence and oversight that the Scottish Government would be able to wield over renewable energy policy. It is clear that energy policy controlled from Holyrood and a national energy regulator for Scotland, alongside control over the CE and other proposals set out in the White Paper on Independence, would facilitate the achievement of Scottish aims and ambitions in this sector. However, taking into account Scottish-specific objectives, whilst remaining both a part of the GB-wide electricity market and the UK-wide CfD FiT mechanism would temper this although the Scottish Government would have more operational control over the CfD FiT mechanism (in terms of setting eligibility rules, subsidy levels and contract allocation) than it currently has. Nonetheless, independence for Scotland under these proposals, then, could have resulted in the situation where *full* control over energy would have remained an elusive dream although the Scottish Government would at least have had a ringside seat.

The independence debate was also characterised by a multitude of claims and counter claims from both sides of the debate. Again, it is worthwhile examining the implications for both governments in the light of the proposals set out in the White Paper on Independence and in order to clarify what this meant for RET deployment. The 100% equivalent Scottish RES-E target will drive continued deployment particularly for RETs and specifically for intermittent wind power. If achieved, this will require a substantial transfer of electricity above levels currently exported outside Scotland, primarily to England and Wales due to the lack of interconnector capacity to other countries. The main issue is that it assumes that the rUK will require electricity generated in Scotland to meet security of supply concerns, renewable energy and climate change targets. The former is given weight by OFGEM's warnings over a looming capacity shortfall in the middle part of this decade. However, there is no guarantee that the rUK will experience a capacity deficit indefinitely whereas independence presumably means just that. Similarly, as things stand, recent analysis by the ENSG, a high-level forum including DECC, OFGEM, the devolved administrations, network companies and generators, highlights that the UK requires a disproportionate contribution of RET capacity of between 6 and 11 GW from Scotland to meet the 2020 RES-E sectoral target (ENSG 2009).

There are a number of additional points of relevance. Firstly, if Scotland voted for independence, the rUK share of the EU 2020 renewables target would most likely be revised down to take this into account, resulting in the rUK requiring less deployment to meet the target. This also assumes that the penalties for not meeting the EU target are stringent enough to ensure full compliance by member states, especially when they have made progress in that direction. Secondly, there is currently no legally binding renewables target post-2020 at the national level. It is difficult to see the UK or the rUK continuing to subsidise renewables beyond 2020, particularly more expensive technology options such as offshore wind and wave and tidal power that Scotland is seeking to develop into a world leading indigenous industry. Thirdly, Scotland is not the only market for RES-E. There is biomass from Austria and Sweden, hydro from Norway and even potentially geothermal from Iceland. Indeed, DECC recently published a consultation looking at opening up the CfD FiT mechanism to generation outside the UK, although if independent this would include Scotland. This point is particularly relevant given the 100% RES-E target: would the rUK need all that expensive generation in comparison to conventional power to meet its decreased EU obligation? And if not, would there be cheaper non-renewable electricity available from elsewhere to meet security of supply concerns? Conversely, the UK would not be the only market for Scottish RES-E generation. The 2009 EU Renewables Directive does allow for statistical transfers that do not require the physical flow of energy from countries in excess of their target commitments. This would bypass the need for expensive interconnectors, although this option should not be ruled out, with progress on the 650 km NorthConnect interconnector between Norway and Scotland and ongoing plans for a EU North Sea super-grid.

The focus has so far been on the implications for the rUK. There are also additional concerns related to the Scottish Government's independence vision. Irrespective of where a RET project is taking place, the pan-UK CfD FiT mechanism is highly complex and the problems of the new mechanism are well documented. As recently as 2012, the House of

Commons Select Committee on Energy and Climate Change announced in an investigation into the EMR proposals that '... *arrangements have become so complex that the proposal has now arguably become unworkable*' (Energy and Climate Change Committee 2012: 4). It is thus a more centralised mechanism than the RO/ROS, and one in which the Scottish Government has lost most operational control over. How much influence would the Scottish Government be able to exert over the mechanism via the proposed partnership, and would this result in a redesign of the mechanism to favour Scottish aims? DECC have already carried out a consultation exploring the possibility that onshore wind generation on the Scottish islands might receive a higher CfD subsidy level than for the rest of the UK, to offset higher costs of transmission and so forth, although DECC continues to delay any decision despite proposals first being put forward three years ago. The '*Scotland's Future*' document also suggested that such a partnership would provide a way for Scottish consumers to avoid paying for new rUK nuclear power whilst keeping the UK-wide renewables subsidy intact. It is difficult to see such an agreement being reached.

Furthermore, there are problems of market participation, route to market and liquidity. This benefits the 'Big Six' vertically integrated energy companies that dominate the UK energy sector, but will particularly impact on small and independent generators. It is doubtful whether any increased powers of oversight for an independent Scotland will enable it to change the way the CfD FiT and the EMR process fundamentally operates. Of importance, the initial report of the 'Delivering Renewable Energy Under Devolution' project suggests that there is no fundamental disagreement about energy in general and renewable energy in particular between the Westminster and the devolved administrations with regard to the prevailing pathway: '*The evidence of our research suggests that... both Scottish and Welsh Governments are broadly comfortable with an energy development pathway that consists of large developments, international investment and conventional generation technologies*' (Cowell et al. 2013: 3).

It is unlikely that independence would have resulted in the prospect of total Scottish control over its energy policy or the achievement of the majority of the aims and objectives as set out in the White Paper on Independence. This in itself should not be surprising, nor should it be taken as evidence of a failing by the Scottish Government given the high degree of shared infrastructure, targets and responsibilities between the

UK and Scotland. As Keay (2013: 2) states, there was always the need for the Scottish Government to tread a '*delicate line between the themes of continuity and change, in relation to energy policy, as in other areas*' in order to reduce uncertainty and risk whilst offering true change to voters despite the difficulties inherent in adopting such an approach. This leads, however, to a key distinction between a vote for independence and a vote to remain in the UK. What role and how much influence will an independent Scottish Government have in any subsequent negotiations? This is important. With the major political parties having offered substantial new devolved powers, this opened up a new space to negotiate irrespective of the referendum outcome. The difference is that if Scotland's constitutional position remains unchanged, as it has, then there is no real guarantee of the further devolution of powers or the type and nature of the powers or in the potential outcomes of any such debate. In other words, although some form of negotiation will occur, it will not be the same process as those taking place in an independent Scotland. As we will see, this is exactly what has happened.

9.3 The Smith Commission Report and the Scotland Act 2016: Further Devolved Powers

The Scottish Government set out its post-referendum proposals in the document More Powers for the Scottish Parliament to inform the Smith Commission into further devolution. As with the pre-referendum White Paper on Independence, it called for increased responsibility over energy and relevant economic policy and an independent Scottish regulator in order to '*deliver a streamlined approach and allow Scotland to design a regulatory and fiscal landscape which maximises the return from the energy sector, encouraging a sustainable industry for the benefit of the people of Scotland*' (Scottish Government 2014a, b: 19). Both the White Paper on Independence and More Powers for the Scottish Parliament also arguably aim for a more cohesive and unified approach to Scottish Government control over Scottish-specific energy and renewable electricity concerns.

The similarities do not end there. The proposals also state that Scotland would continue to participate in the existing GB-wide market for electricity under the single Transmission Operator, National Grid and retention of the CfD FiT (Wood 2014). With Scotland remaining within the UK, these arrangements will undoubtedly continue: further devolution was never likely to change this. Indeed, they would have continued largely unchanged irrespective of the outcome of the independence referendum (Wood 2014). The areas in which the pre- and post-independence proposals offer the prospect of greatest change again lie primarily in the role and influence of the Scottish Government in the overall UK approach to the governance of renewable electricity: the joint oversight of UK-wide bodies including OFGEM, strongly echoing the proposed Energy Partnership outlined in the White Paper on Independence.

It is clear, however, that the new devolved powers as related to renewable electricity set out in the Smith Report fall significantly short of the Scottish Government's proposals and aspirations, and this is echoed in the Scotland Act 2016. Critically, neither the Act nor the Smith Report appears to offer anything new or substantial. What it does offer is essentially more of the same: 'new' devolved powers that are individualist by nature, largely piecemeal and at times vague. Importantly, the devolution of the CE's economic assets and the revenues generated from these assets, including responsibility for around half the Scottish foreshore and the territorial seabed out to 200 nm (Smith Commission 2014), should result in a fundamental shift in the governance of marine renewable energy projects. Control over the CE's leasing arrangements for offshore wind, wave and tidal stream and the required onshore infrastructure will also enable the more optimal management of Scotland's marine assets through the ability to align such developments with national policy objectives.[1]

On the face of it, this is a significant amendment to the existing devolutionary settlement. Devolution of the CE aims to amend the anomalous situation where devolution resulted in the Scottish Government having the lead strategic role in the development of renewable energy in Scotland, including the onshore and offshore planning regimes and economic development but no influence over the leasing rounds. Additionally, the revenues will remain in Scotland instead of going to the UK Treasury. With the Smith Report initially proposing and the Scottish Government subsequently committing to further devolution of these assets to local authority areas, notably Orkney, Shetland and the Hebrides in terms of receiving the full revenue and having a greater say in how the assets are managed, the revenues could be recycled to boost local economic development and community participation and facilitate the developing renewable energy sector instead of transferring to

Westminster. There are, however, issues of growing concern that affect the rights of the Scottish Government. Although existing CE assets and their management will be devolved to the Scottish Government, the CE will be able to continue to acquire property in Scotland (Scottish Affairs Committee, 2012). With regard to the above, the Scotland Act 2016 also sets out extensive and wide-ranging provisions in the interests of defence or national security that appear to act as constraints on the devolved management of pipelines and electricity infrastructure, with no equivalent controls for the management of CE assets in other parts of the UK.

The Scottish Government is also to be given no substantial new devolved powers over renewable electricity. Instead, the Scottish Government is offered '*... a formal consultative role [in] designing renewables incentives and the strategic priorities set out in the Energy Strategy and Policy Statement to which OFGEM must have due regard*' (Smith Commission 2014: 17). The Scotland Act 2016 defines renewable electricity incentive schemes as meaning the CfD FiT, small-scale FiTs and the RO. However, in setting out that the Scottish Government has to be consulted on with regard to the design and delivery of the large-scale CfD FiT, the Energy Act 2013 already contains statutory provisions to this effect.[2] Furthermore, all three Devolved Administrations have already been either consulted on or involved at each stage of the CfD analysis and decision-making process for the EMR delivery plan, including setting the strike price, primarily through joint staff and ministerial meetings and the formal Devolved Administrations Consultation Group (DECC 2012a). The key point is that it fails to devolve any new powers by which the Scottish Government could specifically tailor the mechanism to align it with their own renewable electricity policy objectives. Regarding the ROS, the UK Government, in the same Energy Act, stripped the Scottish Government of devolved powers over the ROS, including setting subsidy levels different to the rest of the UK and potentially keeping the mechanism operational post-2017 in the event of a hiatus in investment due to the CfD FiT. Regardless, then, this offers nothing new and indeed fails even to replace the recently removed devolved powers over renewable electricity.[3]

The Strategy and Policy Statement (SPS) is also highlighted in the Smith Report, although DECC is yet to finalise the statement despite committing to establishing a statutory SPS in 2012 and carrying out a consultation in 2014. However, it is still worthwhile analysing the consultation document here. The rationale behind the SPS is to clarify the

UK Government's strategic priorities (the policy outcomes) in formulating overall UK energy policy and the corresponding roles and responsibilities of the Government, OFGEM and others involved in implementing the policy by defining the desired policy outcomes (DECC 2014a). Again, the Energy Act 2013 sets out that the UK Government already have to consult the Scottish Government on the SPS when drawing up a draft of the statement prior to its designation[4] and under the procedures for undertaking a review of the statement, although only where the statement is to be left unchanged or withdrawn. In contrast, there is currently no statutory requirement for the UK Government to consult the Scottish Government or any other person where the statement is amended, even to the extent that is fully replaced by new content.[5] Presumably, a 'formal consultative role' will fill these gaps, with the Scottish Government consulted on in the development of the statement and any subsequent amendments. Ultimately, however, the content and direction of the SPS will be at the discretion of the Secretary of State, including formulating energy policy and setting out the policy outcomes to be achieved.[6]

The independence debate is over, for the time being. However, with further devolution set out in the Scotland Act 2016, the debate has moved forward. Importantly, the issue of cost socialisation has been largely removed from the debate (however, see below). Furthermore, the question of whether or not the UK would require excess Scottish power generation also no longer applies. The removal of these key obstacles has strengthened calls for an increased role by the Scottish Government in the management and governance of renewable energy. This is also reinforced by recent capacity shortage warnings for the UK overall issued by the energy regulator (OFGEM 2014). In contrast to the UK, Scotland exports surplus power to the rest of the UK, in particular low carbon energy from renewable and nuclear sources in addition to the other benefits discussed previously in this chapter.

9.4 The Implications of Further Devolved Powers on RET Deployment

In practice, the Scotland Act 2016 results in the outcome that the Scottish Government will remain in a consultative role, a position only as good as the level of influence it can wield in practice. The issue then is how this will impact on Scottish renewable electricity policy objectives and renewable electricity technology deployment. Two inter-related questions need to be kept in mind if these proposals are implemented as part of the package of further devolution. Will the Scottish Government be able to affect change on those issues it believes are disproportionately impacting RET deployment and associated infrastructure? And how realistic would this be in practice? Three areas will be examined to highlight the key issues: (1) Regulation; (2) The CfD FiT; and (3) Policy risk.

9.4.1 Regulation

There is to be no Scottish energy regulator or any formal control over regulatory concerns. The Scottish Government will continue to have control over planning consent for energy infrastructure but lack the ability to consent and licence generating and transmission infrastructure which falls under the remit of OFGEM. The same applies over the allocation of new grid capacity or changing the access and charging regime for transmission and distribution. This is a bizarre state of affairs. Northern Ireland already has fully devolved control over energy policy and an independent energy regulator separate from OFGEM, the Utility Regulator for Electricity, Gas, Water, Northern Ireland (UREGNI 2015),[7] yet energy does not carry the same importance as it does in Scotland. Northern Ireland also does not have the same advantages to offer the rest of the UK in terms of environmental (renewables and low carbon), security of supply or domestic energy reserve benefits. This further highlights the point that there are no fundamental difficulties per se in Scotland having the same powers. In addition to the Northern Irish example, there are European cases where this occurs with both similarities and differences between a hypothesised UK–Scotland context including two or more jurisdictions of varying market sizes (big, small) that retain their own separate policies despite policy divergence (Expert Commission on Energy regulation 2014). Critically, by formalising control and the structures and institutions, this could enhance cooperation and understanding between the two governments and ultimately result in the streamlining and simplification of an already complex UK energy policy landscape. Instead, the reforms as part of further devolution avoids this on the one hand, whilst offering a sop on the other, namely by encouraging the Prime Minister of the UK and the First Minister of Scotland to find solutions to the issue of weak inter-governmental working and the lack of respect (Smith Commission 2014).

Returning to the issue of the difficulties arising from the existing and proposed arrangements, a key example is locational charging. Either way, the Scottish Government has no formal powers to amend or remove locational charging. Under the current devolutionary arrangement, it has only been able to influence the direction of policy towards its objectives with regard to locational charging. This was a long and protracted affair, with OFGEM finally deciding after a number of years to reduce the charges levied on Scottish generators but not to remove them. Whether one agrees with this, it does result in significant costs to renewable generators in Scotland, compared to subsidies offered to those in southern England. This appears contradictory given that the benefits of renewable generation, wherever they are located, are ultimately at the UK level: to meet the 2020 target, to encourage investment, create employment and so forth.

It is also unlikely that the enhanced consultative role in the development of the strategic priorities set out in the energy SPS could increase the Scottish Government's influence over locational charging or other regulatory issues of concern to Scottish-specific energy objectives. Even if the Scottish Government could wield substantial influence over the regulator via the reforms set out in the Scotland Act 2016, there is still the matter of the statutory duties that OFGEM has to adhere to, namely in the interest of UK consumers in terms of affordability, environmental and security of supply concerns. Underpinned by various legislations,[8] it is difficult to see how OFGEM could consider Scottish-specific concerns whilst undertaking its statutory duties.

9.4.2 The CfD FiT

By design, the CfD FiT removes virtually all Scottish Government control over the inclusion or exclusion of technologies, for example, the setting of the strike price or subsidy level and the process of contract allocation for new projects. Yet, setting the subsidy level is one key area where the Scottish Government utilised its previously devolved powers to promote Scottish-specific policy objectives, primarily by supporting immature technologies (wave and tidal), maintaining a consistent and stable investment landscape (hydro, onshore wind) or for environmental and other benefits (certain biomass technologies). This power could also have been used to particular effect for other Scottish Government objectives including supporting community and locally owned developments[9] and new technologies such as floating offshore wind turbines.[10]

With the CfD FiT moving towards bidding based on the lowest cost projects being offered contracts first, locational charging could result in a number of Scottish-based projects effectively being forced out of merit. Although the UK Government appears to be moving towards the establishment of a separate tariff 'band' for onshore wind in Orkney, Shetland and the Hebrides (DECC 2014c), albeit again at a slow pace (arguably similar to the slow pace evidenced by HM Treasury regarding the devolution of the Crown Estate and the SPS by DECC), this 'one size fits all' approach ignores the differences between the islands, the extra cost for mainland onshore wind farms and other RET projects in Scotland, particularly larger-scale offshore wind farms located in more expensive to deploy, deeper waters further from the mainland. This is where either greater control for the Scottish Government in terms of new devolved powers over strike prices or some form of formal active participation whether along the lines of the proposed Energy Partnership or some similar process could have resulted in a more nuanced approach to deployment that understood the differences between Scotland and the rest of the UK.

Lacking active participation in the design and operation of the mechanism, however, it is doubtful that the Scottish Government will be able to exert any real influence to further its policy objectives more than is already the case. Even if the Scottish Government received the powers or was able to influence the UK to provide powers under the CfD FiT to set its own strike prices, experience from Northern Ireland argues for caution due to the lack of socialisation of any associated costs if the subsidy levels are set higher in Northern Ireland than the UK position.

9.4.3 Policy Risk

Policy or political risk is arguably unavoidable. It is inevitable that Scottish renewable electricity policy objectives will be open to external policy risk, whether at the UK, EU or global level. In other words, risk outwith the Scottish Government's direct control or influence that can and does affect investor and developer confidence, increasing the costs and/or reducing the optimal deployment of RETs. The reforms (and proposals) make it virtually impossible for the Scottish Government to insulate its renewable electricity policy objectives from UK policy risk. This is highlighted both by Scotland remaining in the GB-wide energy market and the challenges that need to be addressed that currently face key technology options going forward, in particular offshore wind, wave and tidal power. As discussed in Chap. 2, developing these technologies requires concerted and sustained effort by all stakeholders involved in the sector. Additionally, not all deployment barriers lie within the Scottish Government's jurisdictional control. Importantly, Scotland is not in a good place, constitutionally, to negotiate further powers as it would have been if independent although this in itself would not have guaranteed the ability to obtain all the proposals set out in the White Paper on independence.

9.5 Conclusion

The Scottish Government has neither a comprehensive nor cohesive set of devolved powers over renewable electricity. On the surface, Scotland is rightly viewed as a success story in deploying RET capacity. A detailed analysis of the context and implications of devolution, however, reveals that the existing devolutionary settlement is individualistic, piecemeal and arbitrary in what is devolved and reserved between the Scottish and the UK Governments. With the exception of devolving the Scottish assets and responsibilities of the CE, further devolution as set out in the Scotland Act 2016 (and other proposals) does not alter this. This leads to the conclusion that the new set of devolved powers acts as a constraint to the realisation of Scottish-specific renewable electricity policy objectives, with a potential impact on large-scale RET deployment going forward. Critically, it leads to the sub-optimal management and governance of the renewable electricity sector by the Scottish Government. Furthermore, despite the development of an increasingly confident Scottish renewable policy under devolution, and the important contribution of RET capacity in Scotland to the UK overall, the level of influence held by the Scottish Government is insufficient to effect real change. The nature and scope of the devolved powers reaffirm the view that Scotland should remain as a *consultative party* in the governance and management of UK renewable electricity policy. It has arguably managed to influence the UK policy position in certain areas (e.g. locational charging, a separate tariff for onshore wind in the Scottish Islands), but in general this has been a slow, protracted and painful affair and Scottish objectives have not been fully realised, or at all in some cases. The removal of existing devolved powers is of concern.

One missed opportunity arising from the independence debate after the referendum, and conspicuously missing from the Smith Report, was the need for the UK to take more account of sub-national approaches in recognition of the continuing trend towards devolution in Scotland, Wales and Northern Ireland. Some form of Energy Partnership between the Westminster and the devolved administrations can formalise control and the structures for cooperation, streamlining and simplifying the already complex energy landscape across the UK, allowing the promotion of policies, technologies and regulations appropriate to the context of the different nations and regions of the UK. Furthermore, this approach could embrace policy innovation and experimentation at the sub-national level, with the potential to lead to a more productive model of renewable energy governance. This would align with recent research showing that sub-national authorities are '*not merely implementers of EU and Member State norms,* [but] *are also creators of… policies, whose effect extends beyond their own territory*' (Finck 2014: 443–444). It is past time that the UK wakes up to this realisation before it is too late.

Notes

1. Prior to this, the Crown Estate had no statutory obligation to formally consult with Scottish Ministers or take into account any policy objectives or national priorities. Indeed, after devolution in 1997, the Crown Estate closed its separate operating division in Scotland.
2. The Energy Act 2013, Section 24(1), page 16. This also includes Welsh Ministers, the Department of Enterprise Trade and Investment (Northern Ireland) and various others.
3. Although outwith the scope of this article, this appears to suggest that the Scottish Government will have to be consulted on by the UK Government over changes to the small-scale FiT for renewable deployment with an installed capacity less than 5 MW. The statutory basis for this mechanism did not include provisions for this (cf. Energy Act 2008, Sections 41–43, pages 35–38).
4. Energy Act 2013, Part 5, Section 135(3) and (4), pages 103–104.
5. Energy Act 2013, Part 5, Section 134 (6) and (10), page 103.
6. Energy Act 2013, Part 5, Section 131 (2) (a) and (b) and (3), pages 100–101.
7. Utility Regulator for Electricity, Gas, Water, Northern Ireland was formerly known as the Northern Ireland Authority for Utility Regulation (NIAUR).
8. OFGEM's powers and duties are largely provided for in statute, e.g. the Gas Act 1986, the Electricity Act 1989, the Utilities Act 2000, the Enterprise Act 2002 and various Energy Acts (e.g. 2004, 2008, 2010, 2013).

9. This is arguably all the more important now that the UK Government have decided not to extend the small-scale FiT to projects with an installed capacity of 10 MW (see DECC 2014b).
10. The Scottish Government established a different subsidy technology band for floating or innovative offshore wind generation under the ROS. However, the UK Government has decided not to include this technology as a CfD FiT eligible technology and has already closed the RO/ROS (see DECC 2012b).

References

Cowell, R., et al. 2013. Promoting Renewable Energy in the UK: What Difference has Devolution Made? Initial Findings—23rd January 2013' (online), http://cplan.subsite.cf.ac.uk/cplan/sites/default/files/DREUD-FullReport.pdf. Accessed 22 Sept 2014.

Department for Energy and Climate Change [DECC]. 2012a. Annex E EMR Delivery Plan: Decision-making process for Contracts for Difference and the Capacity Market—November (online) 2012, https://www.gov.uk/government/uploads/system/uploads/attachment_data/file/65639/7081-electricity-market-reform-annex-e.pdf. Accessed 12 Jan 2015.

Department for Energy and Climate Change [DECC]. 2012b. Government Response to the Consultation on Proposals for the Banded Support Under the Renewables Obligation for the period 2013–17 and the Renewables Obligation Order 2012–July 2012 (online), https://www.gov.uk/government/uploads/system/uploads/attachment_data/file/42852/5936-renewables-obligation-consultation-the-government.pdf. Accessed 12 Jan 2015.

Department for Energy and Climate Change [DECC]. 2014a. Strategy and Policy Statement—A consultation on the draft statement (URN 14D/1) August 2014 (online), https://www.gov.uk/government/consultations/strategy-and-policy-statement. Accessed 12 Jan 2015.

Department for Energy and Climate Change [DECC]. 2014b. Government Response to the Consultation on Support for Community Energy Projects Under the Feed-in-Tariffs Scheme—November (online) 2014, https://www.gov.uk/government/uploads/system/uploads/attachment_data/file/374540/Govt_Response_to_community_FITs_consultation_-_FINAL.pdf. Accessed 12 Jan 2015.

Department for Energy and Climate Change [DECC]. 2014c. Scottish Islands Renewables—Update Report (online), https://www.gov.uk/government/uploads/system/uploads/attachment_data/file/267481/SIR_update_report_final_template.pdf. Accessed 12 Jan 2015.

Electricity Networks Strategy Group [ENSG], ENSG: Our Electricity Transmission Network: A Vision for 2020—Full Report' (online), http://www.ensg.gov.uk/assets/ensg_transmission_pwg_full_report_final_issue_1.pdf. Accessed 27 July 2014.

Energy and Climate Change Committee. 2012. *Draft Energy Bill: Pre-legislative Scrutiny* (HC 2012–13, 275 I) (online), http://www.publications.parliament.uk/pa/cm201213/cmselect/cmenergy/275/275.pdf. Accessed 2 Dec 2014.

Expert Commission on Energy regulation. 2014. *Energy Regulation in an Independent Scotland* (online), http://www.scotland.gov.uk/resource/0045/00455402.pdf. Accessed 5 Dec 2014.

Finck, M. 2014. Above and Below the Surface: The Status of Sub-National Authorities in EU Climate Change Regulation. *Journal of Environmental Law* 26: 443–472.

Harvie, C. 1995. *Fools Gold: The Story of North Sea Oil*, New ed. London: Penguin Books.

HM Government. 2014. Scotland Analysis: Energy—April 2014 (online), https://www.gov.uk/government/uploads/system/uploads/attachment_data/file/301772/2901910__ScotlandAnalysis_Energy_acc.pdf. Accessed 12 Dec 2014.

Keay, M. 2013. Energy Policy for an Independent Scotland: Continuity or Change? (online), http://www.oxfordenergy.org/wpcms/wp-content/uploads/2013/12/Energy-Policy-for-an-Independent-Scotland.pdf. Accessed 12 Jan 2015.

Office for Gas and Electricity Markets [OFGEM]. 2014. Electricity Capacity Assessment Report 2014 (online), https://www.ofgem.gov.uk/ofgem-publications/88523/electricitycapacityassessment2014-fullreportfinalforpublication.pdf. Accessed 12 Jan 2015.

Pearson, P. and J. Watson. 2010. UK Energy Policy 1980–2010: A history and lessons to be Learnt—A review to mark 30 years of the Parliamentary Group for Energy Studies (online), http://sro.sussex.ac.uk/38852/1/uk-energy-policy.pdf. Accessed 5 July 2014.

Scottish Affairs Committee. 2012. The Crown Estate in Scotland (online), http://www.publications.parliament.uk/pa/cm201012/cmselect/cmscotaf/1117/1117.pdf. Accessed 12 Jan 2015.

Scottish Government. 2008. Energy Policy: An Overview—September 2008 (online), http://www.scotland.gov.uk/Resource/Doc/237670/0065265.pdf. Accessed 15 July 2014.

Scottish Government. 2014a. Scotland's Future: Your Guide to an Independent Scotland—November 2013 (online), http://www.scotland.gov.uk/resource/0043/00439021.pdf. Accessed 27 Nov 2014.

Scottish Government 2014b. More Powers for the Scottish Parliament: Scottish Government Proposals—October 2014 (online), http://www.scotland.gov.uk/Resource/0046/00460563.pdf. Accessed 25 Oct 2014.

Smith Commission. 2014. Report of the Smith Commission for Further Devolution of Powers to the Scottish Parliament—27 November 2014 (online), http://www.smith-commission.scot/wp-content/uploads/2014/11/The_Smith_Commission_Report-1.pdf. Access 27 Nov 2014.

Utility Regulator of Electricity Gas and Water. 2015. About Us: Who We Are (online), http://www.uregni.gov.uk/about_us/. Accessed 12 Jan 2015.

Wood, G. 2014. The Independence Debate and Scottish Renewable Electricity Policy and Practice: Is the referendum that important? September 2014 (online), http://www.scottishconstitutionalfutures.org/OpinionandAnalysis/ViewBlogPost/tabid/1767/articleType/ArticleView/articleId/4217/Geoffrey-Wood-The-Independence-Debate-and-Scottish-Renewable-Energy-Policy-and-Practice-Is-the-Referendum-That-Important.aspx. Accessed 12 Jan 2015.

Author Biography

Dr. Geoffrey Wood is a Teaching Fellow in International Energy Law and Policy at Stirling University. An energy and environmental law, policy and regulatory specialist, his work focuses on the development of legal and governance frameworks for sustainable energy transitions to optimise policy delivery in terms of environmental, social, economic, political and technological outcomes. He has published over 30 articles, books, chapters and other publications. Currently also working as a Researcher and Contributing Lecturer at Dundee University, Geoffrey has previously held research, lecturing and consultancy positions for various organisations, including the Scottish Government and the Offshore Renewables Institute.

CHAPTER 10

Scottish Electricity and Independence

David Toke

10.1 Introduction

This chapter assesses the potential impact of independence on Scottish electricity in the case of independence. Independence is still very much a live issue. Although the referendum saw the independence proposition defeated by around 55–45% in September 2014, the issue refuses to subside. The SNP has greatly increased its parliamentary representation at the UK level, winning 56 out of 59 seats in the 2015 General Election (BBC News 2015) and winning in the Scottish Parliament in 2016 (BBC News 2016). However, the vote to leave the EU in the 2016 referendum ("Brexit") has renewed calls for a further Scottish independence referendum with the Scottish Government publishing draft legislation. Once again, another referendum is back on the table and certainly appears quite plausible. The key question answered here is what are the prospects in continuity and/or change in terms of policy outcomes for Scottish electricity.

It should be said at the start that of course a lot of this is crystal ball gazing. But what we can do here is isolate the plausible from the less plausible. I am not trying to favour or oppose independence in this

D. Toke (✉)
Department of Politics and International Relations,
University of Aberdeen, Aberdeen, UK
e-mail: d.toke@abdn.ac.uk

G. Wood and K. Baker (eds.), *A Critical Review of Scottish Renewable and Low Carbon Energy Policy*, Energy, Climate and the Environment,
DOI 10.1007/978-3-319-56898-0_10

chapter. However, a useful starting point may be to briefly describe these (rather differing) narratives. Then, we can isolate some of the points of divergence and look at how plausible the differing interpretations actually are.

10.2 Referendum Narratives

Perhaps we should begin with the Westminster Government's referendum narrative. This was summarised during the referendum debate by the following passage taken from an article in The Scotsman:

> The report published yesterday says a Yes vote will add at least GBP38 a year to the average household energy bill, possibly rising to as much as GBP189 once the cost of supporting renewables projects is included.
>
> Mr Davey said the existing integrated UK energy market "could not survive" in the event of a Yes vote. The rUK, with a range of power sources domestically and growing grid links with continental Europe, would not need to purchase energy from Scotland.
>
> He added: "For the continuing United Kingdom, the energy relationship with an independent Scotland would become purely commercial - the future of European energy is a single competitive market. We already import more energy to England and Wales from France and the Netherlands than we do from Scotland."
>
> Ewing, (the Scottish Government Energy Minister) added: "Only a Westminster politician could fail to see the huge benefits of Scotland's abundant energy wealth to consumers across these islands. Instead of accessing Scotland's reliable energy resources, he is talking of importing energy over interconnectors that don't yet exist from the European mainland where many countries face a similar energy supply concern as the UK."
>
> Professor Peter Strachan … was among a group of academics who produced a report into the coalition government's decision to fund a new nuclear plant.
>
> He added: "What we've found in the various scenarios that we've presented … [is] that a Scottish Government committed to no nuclear power would actually see a reduction in … consumer bills as a result of having no new nuclear build" (McNab 2014).

On the other hand, an alternative narrative was put out by a report published by the Expert Commission on Energy Regulation established by the Scottish Government (Scottish Government 2014a). This was summarised in the following passage taken from an article in The Scottish Herald:

> An independent Scotland could establish a single energy market with the rest of the UK, provided there was "goodwill and co-operation" between the two countries, according to a report commissioned by the Scottish Government out today.
>
> The Government's independent Expert Commission on Energy Regulation said combined energy markets, built on partnerships between separate countries, existed elsewhere in Europe.
>
> It concluded that a continuing Britain-wide energy market - which the SNP insist would follow independence - would be the best outcome for consumers and utility companies in Scotland, England and Wales.
>
> It also said the rest of the UK should continue to subsidise Scotland's renewable energy sector if the country became independent.
>
> The view contradicts warnings from the UK Government, which has said it would not subsidise green power produced in an independent Scotland and would only buy it "on a commercial basis" (Gardham 2014).

10.3 What Are Main Points of Divergence Here?

First, the 'Yes' and 'No' campaigns differed on the extent to which, or ndeed whether, consumer electricity prices would move upwards under independence. The main focus of this disagreement was who would pay for the incentives paid to renewable energy generators for (a) existing projects and (b) projects constructed after independence. A related issue is the extent to which Scotland could continue to afford to expand renewable energy to achieve its target of 100% of the equivalent of Scottish electricity demand being provided by renewable electricity by 2020.

Second, the opposing campaigns disagreed on the relative likelihood of continued cooperation between Scotland and the rest of the UK, and specifically whether there was a reasonable certainly of a jointly managed transmission system. The 'Yes' campaign assumed that common management of the transmission system would continue, but the UK Government was sceptical about such prospects (DECC 2014a).

In sorting out the accuracy of the competing claims, I want to frame the discussion mainly in terms of how these arguments would affect Scottish independence achieved in the 2020s. There is little purchase in trying to sort out a counterfactual about what would have happened in the event of a 'Yes' vote in 2014. So, for the purposes of the discussion, I will assume that independence occurs in or around the year 2023. I will talk about the issues as they relate to (a) paying for and potentially expanding provision for renewable energy after this date and (b) the extent to which Scotland and the rest of the UK would continue to have joint management of transmission and electricity trading and the extent to which Scotland's electricity system would become more independent of what it is now.

10.4 Renewable Energy After 2020

The prospects for achieving more or less the 100% renewable energy target for Scotland round about 2020 are very good. By the end of 2014, there was enough renewable energy installed for just over half of Scottish electricity to be supplied by RE. 90% of the 2020 target can be achieved if all the currently consented onshore and the offshore wind farms so far (April 2015) given premium price CfD FiTs are implemented. Onshore wind will provide around half of this target, with offshore wind, biomass sources, hydroelectric schemes and solar PV providing the rest.

The issue of who would pay for the renewable energy was hotly contested during the referendum campaign. The Scottish Government argued that under the terms of the 2009 EU Directive then the Westminster Government would have to pay for the achievement of the Scottish renewable energy target on the grounds that this was necessary to achieve the UK Government's commitments. Besides which, England, argued the Scottish Government, needed the power supply that came with the renewable energy (Scottish Government 2013).

The previous Department of Energy and Climate Change refused this argument saying that it could derive the renewable energy from elsewhere, including English-based offshore wind farms. Perhaps most controversially of all, DECC implied that Westminster would stop incentive payments being made in respect of renewable schemes that had already been installed in Scotland before independence. Of course, the longer after 2020 that independence occurred, then the less of a potential issue this could be. The RO, which is the incentive scheme that is being terminated for new schemes in 2016 rather than the previously scheduled 2017 date, will become less burdensome on electricity bills from 2027 as the 20-year funding period of schemes ends. The UK Government is now issuing CfDs for renewable energy. Because these are contracts between the UK Government and the developer, the value of these contracts should not be affected by independence. Of course, the UK Government could decide not to pay the premium price payments for schemes in Scotland, but because this would doubtless be challenged in court and the Government compelled to pay, we can assume that the Government would not try to stop paying people. Under law, a contract states that anyone can sue for breach of contract whatever country they come from.

Under the RO contracting is a private matter, and there is no Government assurance of specific payments to developers, merely a general system of incentivising renewable energy. However, two important points need to be made here. First, the Government would have to change the law to stop incentives under the RO being paid for schemes in Scotland. This is likely to trigger legal action from companies that suffer from this against the government on the grounds of "retrospective" action, which is contra British and European law. But equally, if not more important, if the action by the Westminster Government was going to be aimed at making Scottish rather than English consumers pay for the renewable energy, then it is very debatable about whether this would be the outcome. The people who would suffer would be the electricity companies and others who own the renewable energy schemes. The biggest losers, therefore, would be the two major energy companies, Scottish Power (Iberdrola) and Scottish and Southern Energy who own a high proportion of the schemes in Scotland. They would be unable to recover the losses from consumers under the British electricity market since if they did put up their prices consumers would simply opt to buy from other suppliers. There would be little incentive for the Scottish Government to do this, and they would most likely support the electricity companies in taking action against the Westminster Government. The only way that the Westminster Government could try and make Scottish consumers pay would involve effectively ending the British electricity trading network. That in itself could cause problems for consumers in the rest of the UK. This is especially the case, as it is, where the electricity system is trying to increase its security of supply (Office for Gas and Electricity Markets (OFGEM), 2013; Scottish Government, 2014b). If the British electricity trading system was ended and grid cooperation ceased (one would most likely follow the other) then the rest of the UK would have problems dealing with the variation in renewable energy supplies coming from Scotland. Under European rules, the National Grid could not stop the excess renewable energy coming from Scotland.

In the end, the electricity industry would demand an urgent settlement to any disputes about paying for renewable energy, which is most likely to reflect the status quo of agreed payments. English consumers may not like paying for renewable energy in Scotland if it was independent, but they would probably dislike potential disruption that would come with serious action to enforce this. I will say more about possibilities for joint management of electricity transmission and trading in the next section.

Regardless of whether the rest of the UK would continue to pay incentives to renewable energy schemes installed in Scotland prior to independence, there is still the issue of how continued investments in renewable energy could be paid for after independence. Of course, it could be argued that Scotland might not want to generate more after achieving its 100% target. However, as will be discussed later, Scotland would still be able to sell the electricity generated by renewable energy south of the border at wholesale electricity prices. But Scotland would still have to organise the payment of the incentives and long-term contracts necessary to do this. Such incentives may need to be lower than they are now for technologies such as onshore and offshore wind and solar PV, but it would be optimistic to think that they could entirely disappear much before 2030. Certainly, capital intensive renewable energy plants would still most likely need long-term contracts to guarantee payment of minimum processes for electricity generated.

It may well be the case that Scotland will have a more limited resource base to pay for renewable energy incentives (Toke et al. 2013a). That is because a given amount of incentives for renewable energy has to be shared out among a number of consumers that is less than a tenth compared to the whole of the UK. This would not make much difference but for the fact that there are relatively more renewable energy resources in Scotland compared to the rest of Britain.

One of the paradoxes of UK politics is that Scotland's large amount of renewable energy resource is bound up with the size of its onshore land mass rather than (just) the seas around Scotland. About a third of the UK land area, Scotland has considerable quantities of onshore wind resource (on top of that installed already) and also a lot of land area where solar PV farms could be developed. Both of these sources are relatively cheap and certainly cheaper than nuclear power appears to be. Onshore wind is being paid around £80 per MWh for 15-year contracts (DECC 2015). The cost of solar farms on the ground is also coming down to this sort of level. On the other hand, even on paper, the contract for Hinkley C nuclear power station (agreed in 2013) is for £92.50 per MWh to be paid for more than twice as long (35 years) and backed by government loan guarantees which onshore wind does not receive. In practice, the cost of Hinkley C would be rather larger than this, as witnessed by the difficulty that the project has experienced in being moved forward as the investors in the project seek further government guarantees (such guarantees having an implicit extra price).

Of course, it could be argued that if the rest of the UK spends a lot of money on nuclear power then that would increase energy, increases which would not apply to Scotland if it was independent. Indeed, I and others have previously set out arguments to that effect (Toke et al. 2013b). However, it now looks likely that the nuclear power programme may not materialise, or at the least it seems unlikely to consist of more than a single development delivered very late, quite possibly with large cost overruns over the projected cost. If it is the case that the UK has to focus on renewable energy rather than nuclear power to achieve its decarbonisation aims, then a given renewable energy target for Scotland would be achieved for a relatively smaller increase in electricity price if Scotland remained within the UK compared to being independent. However, set against this is the problem that we now have a government in Westminster that has been cutting back on incentives for renewable energy.

It might be argued that the UK might be induced to pay incentives for renewable energy in an independent Scotland to be installed. However, there would need to be some compelling reason for this given that people of one country do not normally agree to pay for another country's investments in renewable energy. Perhaps if there was another mandatory EU target for renewables (an update of the EU's 2020 targets), this might be the case. However, the EU targets for 2030 are indicative and voluntary in nature. In addition, even if it was the case that the rest of the UK decided it was necessary to "buy in" more renewable energy by paying for its installation (as opposed to just importing it by paying normal wholesale prices), then it would most probably pay only for cheaper onshore wind or solar PV sources rather than more expensive offshore renewable generation. After all, England and Wales have good resources of offshore wind farms themselves.

10.5 Management of Electricity Systems

A significant issue of debate in the referendum was the extent to which Scotland's electricity system would become separate to that of the rest of the UK. It can be argued that the politics of separation might push the (then) two countries towards separate management in order to ensure their own sovereignty and security of supply. Indeed, in Scotland in 2015, there seemed to be sensitivity about security issues when it came to the issue of whether the Longannet coal-fired power plant could be kept running. But how serious are such issues? Does Scotland itself need to ensure that it has a self-contained system of security of supply, or can it safely rely on a future whereby even in independence it can rely on a jointly managed electricity system?

Certainly, during the referendum debate there seemed to be a division of opinion between the 'Yes' and 'No' campaigns about whether Scotland would have to stand on its own as far as the electricity system is concerned. The UK Government implied that there were political barriers to assuming that there could be a commonly managed electricity system after independence.

Yet, I would argue that the UK Government is wrong to imply that issues of political accountability are a barrier for the maintenance of a common System Operator (SO) with Scotland (DECC 2014b, p. 40 para 1.72). Hence, large sections of the UK Government analysis are, at best, irrelevant. In North America, the Mid-continental SO operates across 13 US states in addition to the Canadian state of Manitoba, all of which have different regulatory regimes. It should be straightforward for the UK and Scottish Governments to agree to issue the same license terms to the National Grid so that the National Grid is able to utilise the same grid codes on both sides of the border. It would benefit England and Wales to enjoy a continued common SO. The UK Government is wrong to imply that Scotland would face constraints payments to stop excessive variable cross-border flows of electricity to England if there was no common SO. Under European Network Transmission System Operator-Electricity (ENTSO-E) rules, cross-border trade cannot be prevented and English- and Welsh-based generators would face being constrained as a result of cross-border flows of variable renewable energy supplies. Transmission charges for Scottish electricity would be considerably less under independence than they are now because of ENTSO-E rules.

The UK Government was also wrong to maintain that Scotland could not rely on being able to sell its electricity south of the border. Indeed, according to ENTSO-E rules, it would be illegal for the rest of the UK to try to prevent trading on the electricity wholesale markets. Of course, what this debate is often confused with is whether the rest of the UK would be obliged to pay for the incentives necessary to install new renewable energy schemes in Scotland. That is a different matter, and I discussed this in the previous section. However, once installed, the trading of electricity across borders is not a matter for governments, it is a commercial issue.

10.6 Conclusion: Technology Futures

Normally, making speculations about technological developments would not be expected to significantly affect Scotland as opposed to other parts of the UK. However, in the instance of the likely future growth in use of electric cars, this may affect Scotland very directly in one sense—that is the oil market. If electric cars make a significant impact on the car market—even if it is only 10% of cars in the world—then this will have a big impact on oil prices.

Oil prices react very sensitively to even quite small changes in demand. A significant shift to electric cars around the world will depress oil prices. This will impact the Scottish tax take under independence. So, in that sense it does not matter how much oil there is physically left in the North Sea, what matters to Scotland is that it can be sold at a high price which delivers large tax revenues. However, this tax take may be under severe threat—the only issue being when this will happen rather than whether it will happen. That being said, the drive for Scottish independence is one that rests, in many peoples' minds, on a desire for self-determination as a principle rather than a particular desire to retain oil money. However, the question does arise of the extent to which renewable energy will be able to compensate for oil revenues that may be lost. It has been hoped that income from offshore renewable, such as wave power, tidal stream power as well as offshore wind, might deliver this income.

Whilst there is good hope that the costs of these renewable energy sources will decline, it seems unlikely that they would ever rival oil as a money spinner or that profits could deliver significant tax takes for Scotland. On the other hand, support for these sources may be based not so much on notions of material gain, but on the basis that they are clean energy sources that will not run out or indeed whose prices are not subject to the volatility that afflicts oil markets.

References

BBC News. 2015. *Election 2015: SNP wins 56 out of 59 seats in Scots landslide* (online) http://www.bbc.co.uk/news/election-2015-scotland-32635871.

BBC News. 2016. *Scotland Election 2016—Results* (online) http://www.bbc.co.uk/news/election/2016/scotland.

DECC. 2014a. *Scotland Analysis: Energy* (online) https://www.gov.uk/government/publications/scotland-analysis-energy.

DECC. 2014b. *Scotland Analysis: Energy—Technical Annex* (online) https://www.gov.uk/government/uploads/system/uploads/attachment_data/file/302173/Tech_Annex.pdf.

DECC. 2015. *Contracts for Difference (CFD) Allocation Round One Outcome* (online) https://www.gov.uk/government/uploads/system/uploads/attachment_data/file/407059/Contracts_for_Difference_-_Auction_Results_-_Official_Statistics.pdf.

Gardham, M. 2014. Maintaining a UK-Wide Energy Market Would Be Best Outcome, Report Concludes, The Herald, July 10th.

McNab, S. 2014. "Energy Bills 'Could Rise by GBP200 after Yes Vote'", The Scotsman, April 10th.

OFGEM. 2013. *Electricity Capacity Assessment Report* (online) https://www.ofgem.gov.uk/ofgem-publications/75232/electricity-capacityassessment-report-2013.pdf.

Scottish Government. 2013. *Scotland's Future—Your Guide to an Independent Scotland*, 293–300 (online) http://www.scotland.gov.uk/Publications/2013/11/9348/downloads#res439021.

Scottish Government. 2014a. *Expert Commission on Energy Regulation—Main Report: Energy Regulation in an Independent Scotland* (online) http://www.gov.scot/Resource/0045/00455402.pdf.

Scottish Government. 2014b. *UK Energy Policy and Scotland's Contribution to Security of Supply* (online) http://www.scotland.gov.uk/Topics/BusinessIndustry/Energy/resources/Publications/SecurityOfSupplyApril2014.

Toke, D., F. Sherry-Brennan, R. Cowell, G. Ellis, and P. Strachan. 2013a. Scotland, Renewable Energy and the Independence Debate: Will Head or Heart Rule the Roost? *Political Quarterly* 84 (1): 61–70.

Toke, D., P. Strachan, F. Sherry-Brennan, R. Cowell, and G. Ellis. 2013b. *Is an Independent Scottish Electricity System Good for Renewable Energy and Scotland? University of Aberdeen/Robert Gordon University* (online) http://issuu.com/therobertgordonuniversity/docs/the_dreud_report_2013.

Author Biography

Dr. David Toke is Reader in Energy Politics at the University of Aberdeen. He coordinates a MSc Programme in Energy Politics and Law. He has almost 50 papers published in refereed journals in different disciplines (including Political Science, Geography, Energy and Public Policy). He has published three single-authored monograph books. David has written numerous influential reports for different NGOs, including a report published by the World Future Council which led to the adoption of feed-in tariffs for smaller renewable energy projects in the UK. He continues to be a leading advocate for effective policies to promote renewable energy, and was a leading contributor to the debate about the role of electricity and renewable energy during the 2014 Referendum Debate.

CHAPTER 11

Epilogue: Scotland Moving Forward

Keith Baker and Geoffrey Wood

This book is the result of one of those projects that never seems to end. It began life 6 years ago when Geoff originally signed up to write it before writing up his Ph.D. thesis got in the way, and since then, it has had to be revised around two referendums and several changes to the contributors, but the end result is a work that, at least at the time of writing, is up to date across all the chapters. It is fair to say that there were times when we did not think we would ever see a final draft, and for that, we are indebted to all our contributors, as well as Dave Elliot, Chloe Fitzsimmons and Rachael Ballard at Palgrave.

What we set out to do, and what we hope we have achieved here, is to present a series of essays from leading Scottish experts which does not shy away from being critical where the evidence clearly points towards the need for more radical solutions, and there are many examples of these presented alongside the criticism.

K. Baker (✉)
School of Engineering and the Built Environment,
Glasgow Caledonian University, Glasgow, UK
e-mail: keith.baker@gcu.ac.uk

G. Wood
School of Law, University of Stirling, Stirling, UK
e-mail: geoffrey.wood@stir.ac.uk

G. Wood and K. Baker (eds.), *A Critical Review of Scottish Renewable and Low Carbon Energy Policy*, Energy, Climate and the Environment,
DOI 10.1007/978-3-319-56898-0_11

And yet, if only by the standards of politics in the UK, we live in radical times. On the 23 June 2016, the majority of voters in England and Wales have cast ballots that may now see the majorities in Scotland and Northern Ireland pulled out of the European Union against their will, and against the will of their devolved governments. Significantly, whilst the UK as a whole voted to leave by only 4%, in Scotland almost two-thirds of the electorate voted to remain, giving the Scottish Government, and the Scottish National Party, a clear mandate to prevent Scotland being withdrawn with the rest of the UK. However, with the Conservative Government at Westminster pushing for a hard exit from the EU and the EU standing firm on making any concessions on freedom of movement and free trade, it may be inevitable that the people of Scotland are invited to go to the polls once more, and many people are predicting that the terms of a hard exit will be enough to reignite the passions that emerged around the first independence referendum, and this time deliver a different result.

As the authors gathered here argue, from the point of view of energy policy, there are many risks here, but there are also many benefits. One simple example of these is nuclear. The Scottish Government, reflecting the views of the majority of the Scottish population, has consistently opposed nuclear power, yet it has to use planning law to stop new nuclear build and can do little to stop the upgrading of the National Grid in ways which favour nuclear and centralised conventional generation. This is directly at odds with the policies of a government that sees the future of large scale generation centred around wind and water, whilst the capacities of other renewables continue to develop. Another example is fracking, which is currently under a moratorium in Scotland but is being pushed aggressively by Westminster. In the end, it may be that, as with other aspects of the independence debate, it will come down to practical matters.

And so practically, the prospect of any new nuclear build seems highly unlikely. Even though the Third National Planning Framework (NPF3) now allows for the extension of the lifetimes of Scotland's two nuclear plants at Hunterston and Torness, the spectre of the ongoing and hugely expensive clean-up of the decommissioned Dounreay site still looms large, and the lightning rod of opposition that is the Faslane nuclear weapons base mean being seen to be pro-nuclear is not a risk many MSPs would be willing to take. As such, the loosening of the terms of NPF3 may merely reflect some uncertainty over the risk of a supply–demand gap arising in the next decade but, of course, if Scotland's renewable capacity continues to ramp up at the rates seen this decade this may in future be explained away as being unnecessarily cautious.

However, the Scottish Government, along with some well-known non-governmental organisations, has historically been less prone to caution about the potential contribution of carbon capture and storage to the reduction of emissions from energy generation. This may well have also been influenced by the prospect of Scottish universities and companies queueing up with new patents to licence to China, but at the start of the decade, the hubris around CCS was palpable. However, with the short-lived pilot plant at Longannet being cancelled as far back as 2011 and the succession of cancellations elsewhere in the world, the Scottish Government must now surely be waking up to the realisation that economically viable CCS is not coming any time soon. And this may well be a lucky escape, given the amount of cheap, accessible coal that remains underground. Similarly, the diminishing profits to be made from the North Sea oil that the nationalist parties would see repatriated to Scotland may also, in the end, serve to drive a case for independence that fundamentally relies on making the country a powerhouse for renewable energy.

However, those of us who are pro-independence need to be careful not to get caught up in the idea of it, being a silver bullet for all our problems. What the contributors here have also pointed to are examples of cases where the Scottish Government has failed purely of its own volition, and without recourse to the common refrain that their hands were tied by a Westminster government that the vast majority of Scottish people did not vote for. Regardless of which way people voted, this national re-engagement with politics, and the renewal of an identity that felt decidedly Scottish, as opposed to 'not English', is something that all parties and politicians should regard as a renewal of public consent to be governed, and not necessarily as an endorsement of their policies.

And so, to bring us back to the present, it is now clear that the Scottish people wish for that consent to be extended to being governed by the European Union, which is yet again putting the Scottish Government on a collision course with a Westminster government that has been caught completely unprepared for governing outside of it.

We live in interesting times.

Authors' Biography

Dr. Keith Baker is a Researcher at the School of Engineering and the Built Environment, Glasgow Caledonian University and a co-founder of the Initiative for Carbon Accounting (ICARB). He specialises in energy in the built environment, with a particular focus on fuel poverty, energy efficiency and sustainable building materials.

Dr. Geoffrey Wood is a Teaching Fellow in International Energy Law and Policy at Stirling University. An energy and environmental law, policy and regulatory specialist, his work focuses on the development of legal and governance frameworks for sustainable energy transitions to optimise policy delivery in terms of environmental, social, economic, political and technological outcomes. He has published over 30 articles, books, chapters and other publications. Currently also working as a Researcher and Contributing Lecturer at Dundee University, Geoffrey has previously held research, lecturing and consultancy positions for various organisations, including the Scottish Government and the Offshore Renewables Institute.

Index

G. Wood and K. Baker (eds.), *A Critical Review of Scottish Renewable and Low Carbon Energy Policy*, Energy, Climate and the Environment,
DOI 10.1007/978-3-319-56898-0

W

Z

The manufacturer's authorised representative in the EU is Springer Nature Customer Service Centre GmbH, Europaplatz 3, 69115 Heidelberg, Germany. If you have any concerns regarding our products, please contact ProductSafety@springernature.com

Printed and bound by CPI Group (UK) Ltd, Croydon, CR0 4YY
30/09/2026
02225216-0004